D1793773

Mechanical Circulatory Support

Wayne E. Richenbacher, MD
University of Iowa Hospitals and Clinics
Division of Cardiothoracic Surgery
Iowa City, Iowa, U.S.A.

LANDES
BIOSCIENCE
AUSTIN, TEXAS
U.S.A.

VADEMECUM
Mechanical Circulatory Support
LANDES BIOSCIENCE
Austin

Please address all inquiries to the Publisher:
Landes Bioscience, 810 S. Church Street, Georgetown, Texas, U.S.A. 78626
Phone: 512/ 863 7762; FAX: 512/ 863 0081

ISBN: 1-57059-530-5

Library of Congress Cataloging-in-Publication Data
Mechanical circulatory support / Wayne E. Richenbacher.
    p. cm.
  "Vademecum."
  Includes bibliographical references and index.
  ISBN 1-57059-530-5 (spiral/pbk.)
  1. Blood--Circulation, Artificial--Instruments Handbooks, manu-
als, etc. 2. Extracorporeal membrane oxygenation Handbooks, manuals,
etc. 3. Heart, Mechanical Hanbooks, manuals, etc. 4. Heart failure--
Treatment Handbooks, manuals, etc. I. Richenbacher, Wayne E.
  [DNLM: 1. Assisted Circulation. WG 168 M4855 1999]
RD598.35.A77M43 1999
617.4'1206--dc21
DNLM/DLC   99-33746
for Library of Congress   CIP

Dedication

To my parents who provided me with the opportunity.

To my wife for her unwavering support.

To my children who make it all worthwhile.

Contents

Editor

Wayne E. Richenbacher, MD
University of Iowa Hospitals and Clinics
Department of Surgery
Division of Cardiothoracic Surgery
Iowa City, Iowa, USA
Chapters 1, 2, 3, 5-25

Contributors

Javier H. Campos, MD
Associate Professor
Cardiac Anesthesia Group
Department of Anesthesia
The University of Iowa Hospitals
   and Clinics
Iowa City, Iowa, USA
Chapter 8

Ralph E. Delius, MD
Associate Professor
Division of Cardiothoracic Surgery
Department of Surgery
University of California
Davis Medical Center
   School of Medicine
Co-director
University of California Davis/Sutter
   Children's Heart Surgery Center
Sacramento, California, USA
Chapter 4

Shawn L. Jensen, RN, BSN
Clinical Coordinator
Division of Cardiothoracic Surgery
Department of Surgery
The University of Iowa Hospitals
   and Clinics
Iowa City, Iowa, USA
Chapter 11

Kelly L. Jones, LISW
Department of Social Services
The University of Iowa Hospitals
   and Clinics
Iowa City, Iowa, USA
Chapter 9

Carolyn J. Laxson, RN, MA
Nurse Specialist
The University of Iowa Hospitals
   and Clinics
Iowa City, Iowa, USA
Chapter 9

Scott D. Niles, CCP
Perfusionist
Department of Surgery
The University of Iowa Hospitals
   and Clinics
Iowa City, Iowa, USA
Chapter 11

Angela M. Otto, RN, BSN
ECMO Coordinator
The University of Iowa Hospitals
   and Clinics
Iowa City, Iowa, USA
Chapter 4

James M. Ploessl, CCP
Chief Perfusionist
Department of Surgery
The University of Iowa Hospitals
  and Clinics
Iowa City, Iowa, USA
Chapter 11

Jane E. Reedy, RN, MSN
Director of Clinical Services
Thoratec Laboratories Corporation
Pleasanton, California, USA
Chapter 12

Sarah C. Seemuth, RN, MSN
REMATCH Coordinator
Division of Cardiothoracic Surgery
Department of Surgery
The University of Iowa Hospitals
  and Clinics
Iowa City, Iowa, USA
Chapter 9

Sara J. Vance, RN, BSN
Nurse Specialist
The University of Iowa Hospitals
  and Clinics
Iowa City, Iowa, USA
Chapter 9

Preface

During the past three decades there has been dramatic progress in the surgical management of patients with acute and chronic heart failure. An important component of surgical therapy, mechanical circulatory support, has progressed from the engineering drawing board to the machine shop, the animal laboratory, through rigorous clinical trials to the clinical arena. Mechanical circulatory support devices are now considered a standard of care for patients with failing hearts. With a number of ventricular assist devices having received Food and Drug Administration approval during the past few years, these blood pumps have now become part of the therapeutic armamentarium available in any hospital that has a cardiac surgery program. Most clinicians have only sporadic exposure to blood pump technology during their training, and many hospitals acquiring these systems have not provided care for patients with circulatory support devices previously. Thus we felt there was an opportunity to present, in one volume, a proven approach to the development of a blood pump program and management protocols for patients with circulatory support devices. This text is deliberately light on theory, but does describe a hands-on, how-to approach to the care and management of patients with assisted circulation. This book is intended to serve as a practical guide for cardiac surgeons, fellows and residents, anesthesiologists, cardiologists, perfusionists, operating room, intensive care and ward nurses, bioengineers, physical therapists; essentially any healthcare provider who may come in contact with a patient receiving mechanical circulatory support.

The need for a handbook of mechanical circulatory support was first recognized during my research fellowship in The Division of Artificial Organs at The Pennsylvania State University. Most of what I know about assisted circulation I learned from Dr. William S. Pierce, an internationally recognized pioneer in mechanical blood pump development and the long time Director of the Division of Artificial Organs at Penn State. Over the years, my patients have benefited greatly from the lessons taught to me by Dr. Pierce. For his guidance and support I remain most grateful. Errors or omissions in the preparation of this text are no one's fault but my own. Recognizing that there are nearly as many ways to perform an operation or care for a patient as there are clinicians involved in the patient's care, I would welcome comments or criticisms from anyone who reads this book.

It should be obvious, but is frequently unstated, that caring for this complex patient population is indeed a team effort. The success of the mechanical circulatory support service at The University of Iowa is a tribute to the skill, dedication and expertise of our perfusionists, anesthesiologists, nurses and clinical coordinators. The typing and editorial assistance of Lisa Allen is greatly appreciated. The drawings in this book, that so clearly demonstrate the principles of assisted circulation, were prepared by a gifted medical illustrator, and perfusionist, Richard Manzer. To all that help care for our patients and have assisted with the preparation of this book, my heartfelt thanks.

Overview of Mechanical Circulatory Support

Wayne E. Richenbacher

INTRODUCTION

The concept of assisted circulation began with the development of cardiopulmonary bypass (CPB), (Table 1.1). As the techniques for extracorporeal circulation utilizing CPB were perfected during the 1950s, the era of open heart surgery began. When CPB was introduced into the clinical arena in 1953 open intracardiac repairs were performed for the first time. Ultimately, attempts were made to utilize CPB for temporary support or replacement of cardiac function. However, the oxygenator, an integral component in the CPB circuit has a large blood contacting surface area which results in a significant blood:biomaterial interaction. Activation of the systemic inflammatory response and the resultant capillary leak syndrome produce profound systemic side effects. Furthermore, attempts to utilize CPB for long-term cardiac support were limited by bleeding associated with the need for full systemic anticoagulation and damage to formed blood elements.

As open heart surgery became more commonplace patients with postoperative ventricular dysfunction were encountered creating a need for a device capable of supporting the patient's circulation.[1] The simplest such device, the intraaortic balloon pump (IABP) is a catheter mounted balloon that is positioned in the descending thoracic aorta. When properly timed with the patient's native cardiac rhythm the IABP is capable of improving coronary perfusion, reducing left ventricular afterload and augmenting cardiac output to a very modest degree. The IABP was first employed clinically in 1968. Milestones in IABP development include the conversion from an open to percutaneous insertion technique in 1980, manufacture of balloons of smaller dimension that can be employed in patients with a small body habitus and in the pediatric patient population and timing schema that permit balloon pump use in patients with rapid and/or irregular cardiac rhythms. The IABP is now a standard form of therapy for patients with a

1

Table 1.1. Timeline for the development of mechanical circulatory support devices

1953	First clinical application of cardiopulmonary bypass.
1963	First clinical use of a left heart assist device.
1968	First clinical application of intraaortic balloon counterpulsation.
1969	First insertion of a pneumatic artificial heart as a bridge to transplantation.
1980	First percutaneous intraaortic balloon pump insertion.
1982	First permanent implantation of a pneumatic artificial heart.
1992	First Food and Drug Administration (FDA) approval of a pneumatic ventricular assist device (VAD) for use in a patient with postcardiotomy cardiogenic shock.
1994	First FDA approval of a pneumatic VAD for use in a patient as a bridge to cardiac transplantation.
1998	Initiation of first FDA approved clinical trial in which an electric VAD is implanted as a permanent form of circulatory support.
1998	First FDA approval of an implantable electric VAD for use in a patient as a bridge to cardiac transplantation—including home discharge.

variety of cardiovascular diseases (see Chapter 3: Intraaortic Balloon Counterpulsation).

There remained a need for a cardiac replacement device.[2] During the 1960s and 1970s a variety of intracorporeal and extracorporeal mechanical blood pumps were developed and tested in the laboratory with occasional clinical use. During the 1980s clinical experience increased. Patient selection criteria and management techniques were developed and refined. Although blood pumps were used in a variety of clinical situations by the end of the 1980s, interim support of the circulation with a mechanical blood pump was proven efficacious in two distinct patient populations: those patients requiring temporary mechanical circulatory support pending ventricular recovery (see Chapter 5: Ventricular Assistance for Postcardiotomy Cardiogenic Shock), and patients who required a mechanical bridge to cardiac transplantation (see Chapter 6: Ventricular Assistance as a Bridge to Cardiac Transplantation). During the past 9 years a number of ventricular assist devices (VADs) have received Food and Drug Administration (FDA) approval. Most recently, the first FDA approved clinical trial in which a VAD is implanted as a permanent form of circulatory support was initiated in the United States in 1998.

DEVELOPMENTAL HURDLES ENCOUNTERED DURING MECHANICAL BLOOD PUMP DEVELOPMENT

Developmental at hurdles for cardiac replacement devices are listed in Table 1.2. Blood contacting surfaces have to be nonthrombogenic in order to avoid thromboembolic complications during the period of mechanical circulatory support. Blood contacting surface design varies from ultrasmooth, seam-free surfaces to textured surfaces that actually promote stable neointimal formation. In general, blood pumps are constructed with flow dynamics that minimize areas of blood

---

Table 1.2. Developmental hurdles for cardiac replacement devices

1. Nonthrombogenic blood contacting surface.
2. Pumping action that avoids blood trauma.
3. Control schema.
4. Size and configuration.
5. Portability.
6. Reliability.

---

stasis within the pump itself. Regardless of whether the blood pump is nonpulsatile, as in the case of a centrifugal pump or pulsatile, the pumping action must be gentle enough to avoid blood trauma. Injury to formed blood elements can result in platelet and white blood cell activation as well as hemolysis. Control systems in the more sophisticated pulsatile devices allow these pumps to function without supervision by a trained healthcare provider. In addition, the pumps are physiologically responsive, raising and lowering the patient's cardiac output in accordance with activity level. In the case of the total artificial heart, the control system must also balance the output of the two prosthetic ventricles. The blood pumps must be capable of generating forward flow comparable to the cardiac output of a native heart. The size of the blood pump and configuration of the cardiac replacement device assumes greater significance in the total artificial heart as it must be located within the patient's pericardium.

Although portable pneumatic drive units are currently the subject of clinical trials, commercially available air driven mechanical blood pumps require a bulky external drive unit. The patients are able to move about, but they frequently require assistance and experience a less than satisfactory quality of life. Portability becomes a more pressing issue in patients who require long-term circulatory support. Electrically powered devices are considerably more complex but require a much smaller controller and power source. The electric motor is implanted as part of the blood pump. The external controller and battery pack can be worn on a belt, satchel or shoulder harness eliminating the need for a wheel based drive unit. Reliability is of critical importance as these devices must pump blood for months or even years without malfunction. An in-depth discussion of these developmental issues is beyond the scope of this handbook. Suffice it to say that the mechanical blood pump industry is tightly regulated.[3] In the United States, the Medical Device Amendment of 1976 amended the Federal Food, Drug and Cosmetic Act to approve clinical investigation of medical devices and approve new medical devices before commercialization. Rigorous preclinical laboratory and animal testing as well as carefully constructed clinical trials require FDA approval before a device can be released for general use.

Two general classes of mechanical blood pumps have been developed. The VAD is essentially half an artificial heart and is designed to replace the function of a single ventricle. Two VADs can be employed for biventricular support. Blood enters and leaves the VAD by way of inlet and outlet cannulae. For right ventricular assistance the inlet cannula is inserted into the right atrium. For left ventricular

1

assistance the inlet cannula may be inserted into either the left atrium or left ventricular apex depending upon VAD type and the patient's disease state. The outlet cannulae are attached or inserted into the main pulmonary artery or ascending aorta depending upon which ventricle the VAD is supporting. Thus the right VAD (RVAD) functions in parallel with the patient's heart while the left VAD (LVAD) may function either in series or parallel with the patient's heart depending upon the cannulation scheme. A variety of VADs are available and the location of the blood pump determines whether the VAD is classified as either an intracorporeal or extracorporeal device (Table 1.3). The intracorporeal ventricular assist systems include a blood pump that is located within a body wall or cavity.[4] The blood pump is connected by a percutaneous drive line to an external controller and power source. Extracorporeal ventricular assist systems employ percutaneous inlet and outlet cannulae. Extracorporeal ventricular assist systems may be further subdivided depending upon the location of the blood pump. Paracorporeal devices employ a blood pump that is positioned adjacent to the patient's abdominal wall. The blood pump is in turn connected by a lengthy drive line to a bedside console containing the controller and power source. An alternative design employed with centrifugal pumps and the Abiomed BVS 5000 Biventricular Support System (Abiomed, Inc., Danvers, MA) incorporates longer inlet and outlet tubing. The blood pump itself is located at the patient's bedside.

As a VAD is not positioned within the patient's pericardium, blood pump design is not limited by size or anatomic constraints. Should the device malfunction the patient's failing heart will frequently function well enough to allow a brief period of time for intervention or repair. When compared to the artificial heart, the VAD is technically easier to implant. CPB time required for implantation is brief and in some instances the VAD can be implanted without the use of CPB. Unlike the IABP, the VAD is capable of completely replacing the function of the native heart. In the face of low pulmonary vascular resistance an LVAD can often maintain a satisfactory cardiac output even in the absence of an effective cardiac rhythm.

The total artificial heart is a complete biventricular replacement device (Fig. 1.1).[5] Although these devices have received the greatest attention in the popular press, they remain investigational. Artificial hearts are the most complex mechanical blood pumps both in design and function. As the artificial heart must fit within the pericardium, anatomic and size considerations are paramount. The sophisticated control system not only varies the device output with the patient's physiologic need but must also balance the output of the two ventricles in order to accommodate the bronchial circulation. Reliability considerations also assume greater importance as there is no "back-up" in the event that the artificial heart should fail.

Table 1.3. Ventricular assist devices

Type/Name	Manufacturer	Blood flow	Power source	Left, right, or biventricular support	Blood pump location
Centrifugal*	Several	Nonpulsatile	Vortex or impeller pump	L, R, Bi	Extracorporeal
BVS 5000 Biventricular Support System	Abiomed, Inc.	Pulsatile	Pneumatic	L, R, Bi	Extracorporeal
Thoratec VAD System	Thoratec Laboratories Corp.	Pulsatile	Pneumatic	L, R, Bi	Paracorporeal
HeartMate IP LVAS	Thermo Cardiosystems, Inc.	Pulsatile	Pneumatic	L	Intracorporeal
HeartMate 1000 VE LVAS	Thermo Cardiosystems, Inc.	Pulsatile	Electric	L	Intracorporeal
Novacor N-100 LVAD	Novacor Division, Baxter Healthcare Corp.	Pulsatile	Electric	L	Intracorporeal

*   Centrifugal blood pumps are not approved by the Food and Drug Administration for use as ventricular assist devices. They are, however, included in the discussion of mechanical ventricular assistance as they are frequently employed for short-term circulatory support.
L = left, R = right, Bi = biventricular

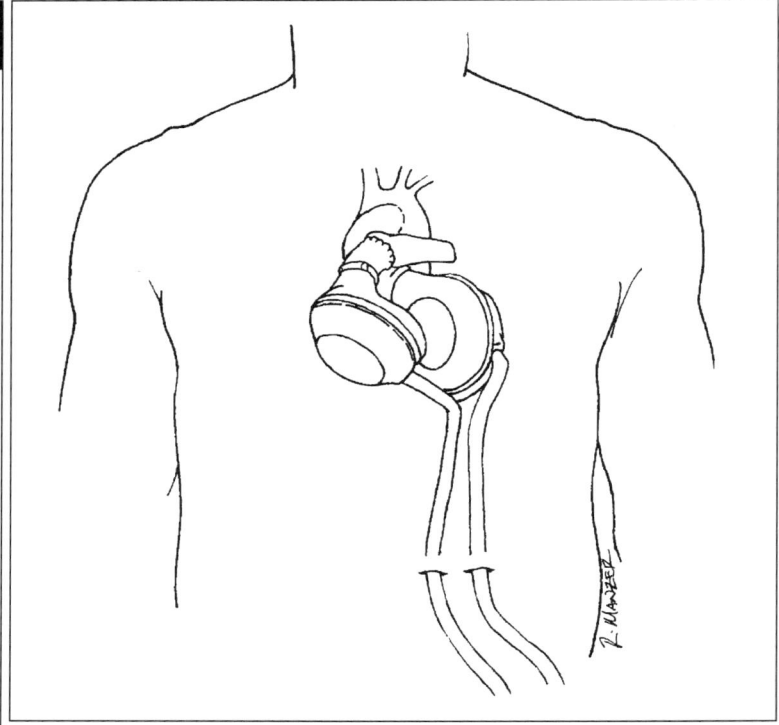

Fig. 1.1. Implantation of the pneumatic artificial heart requires removal of the patient's native heart. Atrial cuffs on the artificial heart are sewn to the patient's atrial remnants in a manner similar to the Shumway technique for cardiac transplantation. The outlet grafts are sewn to the appropriate great vessel.

MANAGEMENT OF HEART FAILURE

In order to understand the role of mechanical circulatory support in the continuum of care provided to patients with heart failure it is helpful to briefly review therapeutic modalities available to heart failure patients (Table 1.4). The medical management of heart failure has improved dramatically in recent years with the introduction of beta-blockers and angiotensin-converting enzyme inhibitors.[6,7] Although somewhat more controversial, with little literature evidence to support an increase in survival, chronic home inotrope infusion therapy and intermittent outpatient inotrope infusion have been offered to heart failure patients. Surgical modalities available to heart failure patients include high risk myocardial revascularization and valve replacement. Coronary artery bypass is increasingly offered

---

Table 1.4. Therapies available to heart failure patients

Medical
    Beta-blockade
    Angiotensin-converting enzyme (ACE) inhibitors
    Diuretic therapy
    Outpatient/home inotrope infusion therapy
Surgical
    Targeted coronary artery bypass
    Cardiac transplantation
    Cardiomyoplasty
    Ventriculectomy

---

to patients with low ejection fractions who have evidence of viable myocardium based upon thallium scans and positron emission tomography. Cardiac transplantation, the mainstay of surgical care for patients with end stage cardiomyopathy is limited by a persistent shortage of donor hearts and limited long-term survival in patients who undergo cardiac transplantation secondary to complications of immunosuppressive therapy and allograft coronary artery disease.[8] Cardiomyoplasty[9] and ventriculectomy[10] have generated a great deal of interest in the recent past but their efficacy is largely unproven.

Current indications for mechanical circulatory support with a VAD involve two patient groups. A patient with refractory cardiogenic shock following an open heart operation may be a candidate to receive days to a week or two of ventricular assistance (see Chapter 5: Ventricular Assistance for Postcardiotomy Cardiogenic Shock). The expectation in this clinical application is that ventricular recovery will occur at which time the blood pump is removed. Patients who are approved candidates for cardiac transplantation but who fail hemodynamically prior to the availability of a donor heart may receive a VAD as a mechanical bridge to transplantation (see Chapter 6: Ventricular Assistance as a Bridge to Cardiac Transplantation). In this scenario the VAD remains in place for months to a year or more and is only removed at the time of recipient cardiectomy and cardiac transplantation.

As heart failure is such a prominent health problem in our aging population we can expect this healthcare field to continue to rapidly evolve. On the horizon, skeletal muscle ventricles may be used to augment native cardiac output. Xenotransplantation, in which animal hearts are used for cardiac transplantation, is a promising field that has been hampered by seemingly insurmountable immunologic hurdles.[11] The development of gene transfer techniques to genetically alter animal endothelium holds great promise for the future. Aggressive alternative medical therapies for heart failure include medicines that alter the neurohumoral milieu, including dopamine-hydroxylase inhibitors and drugs that reduce sympathetic tone.

1

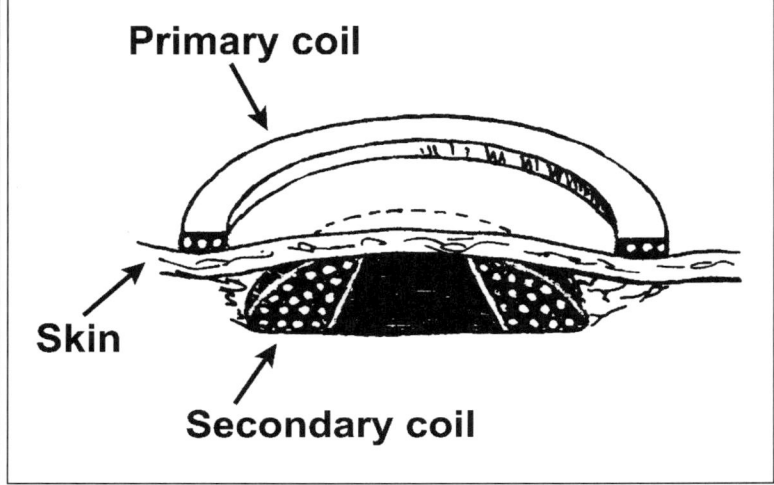

Primary coil

Skin

Secondary coil

Fig. 1.2. Transcutaneous energy transmission utilizes an external primary coil and internal secondary coil with inductive coupling to pass energy across intact skin. When electric ventricular assist devices employ transcutaneous energy transmission, the need for a percutaneous drive line will be eliminated.

FUTURE OF MECHANICAL CIRCULATORY SUPPORT

The ultimate goal of blood pump development is to construct a completely implantable device designed for permanent use.[12] Technological hurdles yet to be overcome include the elimination of the percutaneous drive line, development of an implantable control system and power source to permit brief periods of tether-free operation and construction of a reliable compliance system for use with implantable pulsatile VAD systems. The percutaneous drive line serves as a potential ongoing source of infection. Investigational, completely implantable electric ventricular assist systems employ transcutaneous energy transmission techniques where energy is passed across the intact skin using internal and external coils and transformer technology (Fig. 1.2). These completely implantable VAD systems incorporate an implantable controller and back-up battery that allows the external primary power source to be removed for short periods of time. Higher energy density batteries will ultimately allow longer periods of tether-free function which will further improve the patient's quality of life. Lastly, even in electrically powered VADs air must move in and out of the blood pump as the blood chamber fills and empties. Currently, implantable devices employ a percutaneous vent as part of the drive line. When the drive line is eliminated and the completely implant-

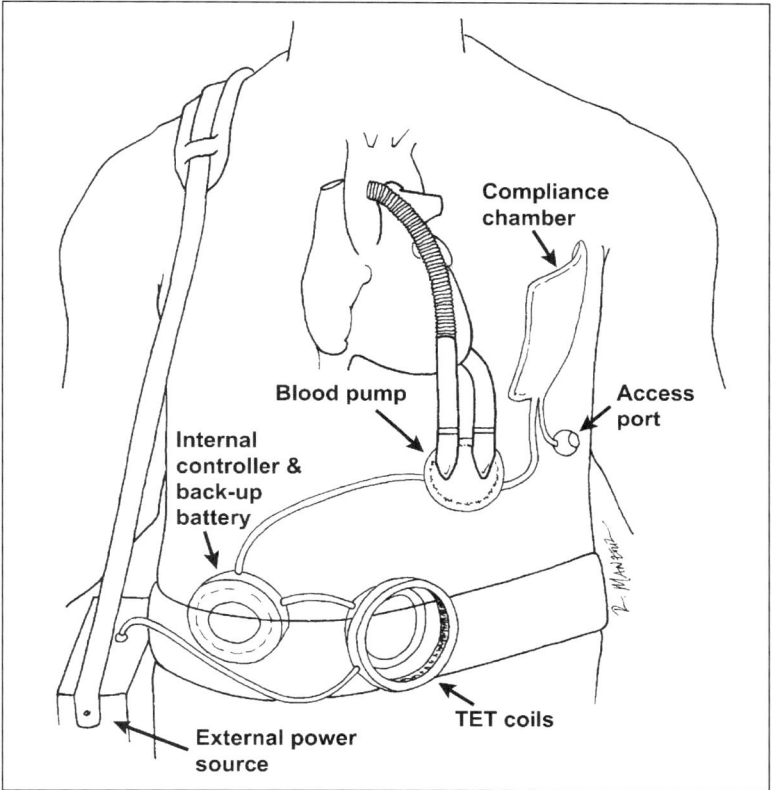

Fig. 1.3. The completely implantable ventricular assist system. The sealed system will employ transcutaneous energy transmission (TET) coils, an implanted controller and back-up battery and compliance chamber.

able blood pump is sealed air, displaced from the blood pump during VAD diastole will be handled by an implanted compliance chamber (Fig. 1.3). Problems encountered during compliance chamber development include loss of compliance over time secondary to pseudocapsule formation and a slow mass transfer of gas across the biomaterial utilized in the compliance system. A newer generation of compact, electrically powered axial flow pumps are also in the later stages of laboratory development. These devices are nonpulsatile, a design feature that eliminates the need for a compliance chamber altogether.[13] Clinical trials involving completely implantable VAD systems, both pulsatile and nonpulsatile, are anticipated within the next few years.

REFERENCES

1. Argenziano M, Oz MC, Rose EA. The continuing evolution of mechanical ventricular assistance. Current Problems in Surgery 1997; 34:317-388.

2. Richenbacher WE, Pierce WS. Key references: Mechanical circulatory support. Ann Thorac Surg 1996; 62:1558-1559.

3. Hill JD. How much do we need to know before approving a ventricular assist device? Ann Thorac Surg 1993; 55:314-328.

4. Goldstein DJ, Oz MC, Rose EA. Implantable left ventricular assist devices. New Engl J Med 1998; 339:1522-1533.

5. Copeland JG, Arabia FA, Banchy ME et al. The CardioWest total artificial heart bridge to transplantation: 1993 to 1996 national trial. Ann Thorac Surg 1998; 66:1662-1669.

6. Massin EK. Outpatient management of congestive heart failure. Tex Heart Inst J 1998; 25:238-250.

7. Delgado R, Massin EK, Cooper JR. Acute heart failure. Tex Heart Inst J 1998; 25:255-259.

8. Hauptman PJ, O'Connor KJ. Procurement and allocation of solid organs for transplantation. New Engl J Med 1997; 336:422-431.

9. Furnary AP, Jessup M, Moreira LFP. Multicenter trial of dynamic cardiomyoplasty for chronic heart failure. J Am Coll Cardiol 1996; 28:1175-1180.

10. Batista RJV, Verde J, Nery P et al. Partial left ventriculectomy to treat end-stage heart disease. Ann Thorac Surg 1997; 64:634-638.

11. DiSesa VJ. Cardiac xenotransplantation. Ann Thorac Surg 1997; 64:1858-1865.

12. Sapirstein JS, Pae WE Jr, Rosenberg G et al. The development of permanent circulatory support systems. Semin Thorac Cardiov Surg 1994; 6:188-194.

13. Westaby S, Katsumata T, Evans R et al. The Jarvik 2000 Oxford system: Increasing the scope of mechanical circulatory support. J Thorac Cardiovasc Surg 1997; 114:467-474.

Developing a Mechanical Blood Pump Program

Wayne E. Richenbacher

MECHANICAL CIRCULATORY SUPPORT TEAM

A mechanical circulatory support team is an integral part of a heart failure treatment program. Although a formal heart failure treatment program is only necessary at cardiac transplant centers, nontransplant centers should also create a multidisciplinary team capable of providing specialized care to patients requiring mechanical ventricular assistance. Representative members of a mechanical circulatory support team are listed in Table 2.1. The cardiothoracic surgeon should be trained in all aspects of device usage including patient selection, device implantation and perioperative management. Heart failure cardiologists receive variable exposure to mechanical blood pump technology during their training. Thus, it is important that cardiologists who have not previously been exposed to this technology be made aware of the role of mechanical ventricular assistance in patients with acute and chronic heart failure. Academic institutions that are active in blood pump development may include biomedical engineers on the mechanical circulatory support team. Most institutions, however, rely upon perfusionists or specially trained nurses to assist with device implantation and the day-to-day management of the blood pump drive unit. Routine open heart operations usually require the services of one perfusionist who runs the cardiopulmonary bypass (CPB) machine. When a mechanical blood pump is to be implanted the services of a minimum of two perfusionists are required. One perfusionist is responsible for the CPB machine while the second perfusionist manages the ventricular assist device (VAD) drive unit. The presence of an intraaortic balloon pump (IABP) console may necessitate additional technical support personnel.

Highly motivated nursing staff should be identified and invited to serve as members of the mechanical circulatory support team.[1] Nurses from all patient care areas should be included. Operating room nurses are specifically trained in

---

2

Table 2.1. Mechanical circulatory support team

Cardiothoracic Surgeon
Heart Failure Cardiologist
Clinical Coordinator
Perfusionist
Nursing Staff
       Operating Room
       Intensive Care Unit
       General Ward
Cardiac Anesthesiologist
Consultants
       Nephrologist
       Neurologist
       Infectious Disease
       Hematologist/Blood Bank
       Psychiatrist/Psychologist
Social Worker
Cardiac Rehabilitation/Physical Therapist
Financial/Billing Counselor
Dietitian
Ethicist/Pastoral Care Representative

device preparation, assembly and specialized implantation techniques. When the VAD patient arrives in the surgical intensive care unit 2:1 nursing is provided for the first 24-48 hours or until such time as the patient's condition stabilizes. The intensive care unit and ward nurses are trained in drive console management and are prepared to deal with common alarm situations. Primary caregivers must also understand the complex physiology that can occur in a patient receiving mechanical blood pump support. This requires that general ward nursing staff be cross-trained so that they can deal with medical and surgical management issues. Nurses are responsible for wound care, in particular, involving the percutaneous drive line or cannulae exit sites. As such they must be comfortable with sterile technique. The nurses must also be sensitive in dealing with long-term psychosocial issues in patients who may be hospitalized for prolonged periods of time. Critical to the development and ongoing success of a mechanical circulatory support team is the designation of a clinical coordinator. This individual frequently has a nursing background although a perfusionist or bioengineer may function well in this regard. The clinical coordinator serves to prepare the patient and his family for device implantation, ensures that staff and patient training in the management of the device is performed and deals with ongoing medical and device related issues during the period of mechanical circulatory support.

Patients who require mechanical circulatory support frequently manifest unique clinical problems. Thus, it is helpful to identify one or two individuals to serve as consultants from a variety of medical disciplines. These consultants understand the nuances of mechanical blood pump support and the impact of a device on the patient's clinical condition. Their input is frequently invaluable in ongoing patient management. Patients with heart failure present with varying degrees of renal dysfunction. A nephrologist will assist with their management and organize ultrafiltration or dialysis services as required. The blood:biomaterial interface in the mechanical blood pump is potentially thrombogenic. We request that a neurologist evaluate each patient prior to mechanical blood pump insertion. If a thromboembolic event occurs following blood pump insertion the neurologist can usually identify subtle changes in neurologic status. Chronically hospitalized patients who require ongoing invasive hemodynamic monitoring and one or more major cardiothoracic surgical procedures are at risk for the development of nosocomial infection. Specialists in infectious disease can assist with recommendations regarding appropriate antibiotic therapy. Blood usage can, on occasion, be excessive and hematologic derangements related to prolonged CPB time and complex implantation procedures may be difficult to manage. A hematologist or blood bank personnel with specialized expertise in managing a multifactorial coagulopathy can assist the surgical team by recommending and arranging for appropriate blood component therapy. Patients with chronic disease, who are potentially housed in the hospital for prolonged periods of time benefit greatly from counseling.[2] We prophylactically employ the services of a social worker, pastoral care representative and, when necessary, a clinical psychologist. The social worker is also an invaluable resource when dealing with the patient's insurance carrier and place of employment and when assisting with family and patient relocation issues. The social worker is often called upon to work in conjunction with billing office personnel to develop creative strategies when seeking reimbursement for the patient's hospital bill and physician's services (see Chapter 12: Financial Aspects of a Mechanical Circulatory Support Program). Patients in chronic congestive heart failure are often nutritionally depleted. Poor exercise tolerance, chronic deconditioning and associated muscle wasting complicated by early satiety and a poor appetite following VAD insertion are issues most effectively addressed by a licensed dietitian. The dietitian can assess the severity of the patient's cachexia and provide recommendations for nutritional repletion following device implantation. Finally, the critical nature of end stage heart failure frequently leads to difficult life and death decision making. The impartial view of an ethicist complements the clinical decisions made by the patient's healthcare team.

SELECTING A VENTRICULAR ASSIST SYSTEM FOR POSTCARDIOTOMY CARDIOGENIC SHOCK

A mechanical blood pump system should only be utilized at an institution staffed and capable of conducting open heart operations (Table 2.2). The surgeon directing the mechanical blood pump program must first decide upon the indications for use of mechanical circulatory support. Will the device be used for patients in postcardiotomy cardiogenic shock, as a mechanical bridge to cardiac transplantation, or both? In estimating the volume of patients who would require mechanical blood pump support for postcardiotomy cardiogenic shock it has been

2

Table 2.2. Food and Drug Administration approved ventricular assist devices

Device	Application	Date approved	# of centers: US*	# of centers outside US
Abiomed BVS 5000 Biventricular Support System	PCCS	11/20/1992	> 400	> 60
Thoratec VAD System	Bridge PCCS	12/20/1995 5/21/1998	Total 55	Total 39
Thermo Cardiosystems, Inc. HeartMate 1000 IP LVAS	Bridge	10/3/1994	89	33
Thermo Cardiosystems, Inc. HeartMate VE LVAS	Bridge	9/20/1998	25	28
Novacor N-100 LVAD	Bridge	9/28/1998	22	40

* Numbers of centers accurate as of February 1999
PCCS = Postcardiotomy cardiogenic shock
Bridge = Bridge to cardiac transplantation

shown that 0.5-1% of patients who undergo an open heart operation will be difficult to wean from CPB and will require an advanced form of mechanical circulatory support.[3] With improved myocardial protection techniques and a newer generation of inotropes, a conservative estimate is that fewer than 5-10 patients would require a mechanical blood pump for postcardiotomy cardiogenic shock in any given year at the average healthcare center. By developing a realistic estimate of the number of patients who would benefit from mechanical circulatory support, hospital administration with guidance from the clinician can factor budgetary considerations into device selection.

Currently, three ventricular assist systems are used to support patients with postcardiotomy cardiogenic shock (see Chapter 5: Ventricular Assistance for Postcardiotomy Cardiogenic Shock), (Table 2.3). The Abiomed BVS 5000 Biventricular Support System (Abiomed, Inc., Danvers, MA) and Thoratec VAD System (Thoratec Laboratories Corp., Pleasanton, CA) are extracorporeal, pneumatic, pulsatile systems that can be configured for either right, left or biventricular support. These systems are Food and Drug Administration (FDA) approved for this particular clinical application. In general, the Abiomed system is less expensive and arguably simpler to insert due to the small diameter of the cannulae (Figs. 2.1-2.3).[4,5] However, the Abiomed VAD is located at the patient's bedside and filled by gravity; both are features which tend to limit patient mobility. The Thoratec device is more expensive (Figs. 2.4, 2.5). However, the Thoratec VAD is paracorporeal and although the patient is tethered to a sizable drive unit, mobility is enhanced (Fig. 2.6).[6] Furthermore, the Thoratec device is also FDA approved for use as a mechanical bridge to cardiac transplantation. This approval provides

Table 2.3. Ventricular assist systems used for patients with postcardiotomy cardiogenic shock

Device	Advantages	Disadvantages
Centrifugal pump	Readily available. Inexpensive. Most perfusionists, surgeons well versed in its use.	Not FDA approved for use as a VAD. Drive unit requires constant attendance.
Abiomed BVS 5000 Biventricular Support System	FDA approved for use in postcardiotomy cardiogenic shock. Relatively inexpensive. Simple to prime pump and insert cannulae. Drive unit functions independently.	Relatively small inlet cannulae and gravity fill design may limit flow. Patient must remain in intensive care unit. Limited patient mobility.
Thoratec VAD System	FDA approved for use in postcardiotomy cardiogenic shock and as a bridge to cardiac transplantation. Patient may be moved out of intensive care unit. Drive unit functions independently. Increased patient mobility.	Expensive.

FDA = Food and Drug Administration
VAD = Ventricular assist device

greater flexibility in patient management. A patient who is supported with either the Abiomed or Thoratec VAD system at a nontransplant center may be transferred to a cardiac transplant center should the patient not recover ventricular function. If the patient is deemed a suitable candidate for transplantation and is supported with the Abiomed VAD, the device must be removed and a pump designed for long-term use implanted. The need for this secondary procedure is obviated in the patient supported with the Thoratec VAD. If the patient does not recover ventricular function and is found to be a cardiac transplant candidate, the patient can simply be maintained on the Thoratec VAD until a cardiac transplant is performed. The drive units employed with both the Abiomed and Thoratec VAD systems are self-regulating and beyond the immediate postoperative period can be readily managed by trained nursing staff.

A third VAD system that is frequently employed in the postcardiotomy cardiogenic shock patient population is the centrifugal blood pump (Figs. 2.7, 2.8).[7] This class of pumps has not been approved by the FDA for use in this clinical application. These devices are quite inexpensive compared to the sophisticated pulsatile devices described above. They are simple to implant and as they do not

Fig. 2.1. The dual chamber Abiomed BVS 5000 ventricular assist device. (Photograph courtesy of Abiomed, Inc.).

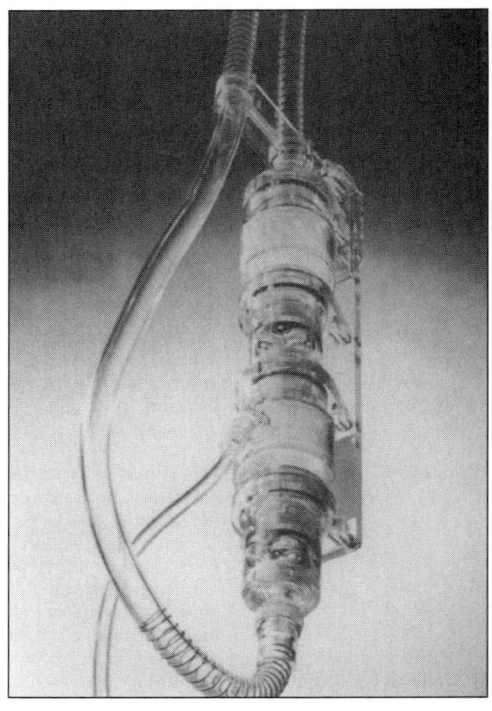

Fig. 2.2. Cross section of the Abiomed BVS 5000 ventricular assist device. A. During blood pump diastole blood moves from the atrial chamber through the inlet valve into the ventricular chamber. B. During blood pump systole a pulse of air from the drive console compresses the blood sac in the ventricular chamber. Blood is ejected through the outlet valve and outlet graft back to the patient. The atrial chamber fills via gravity.

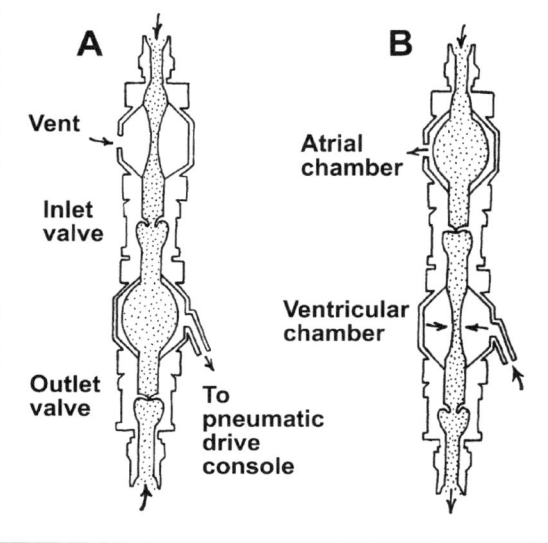

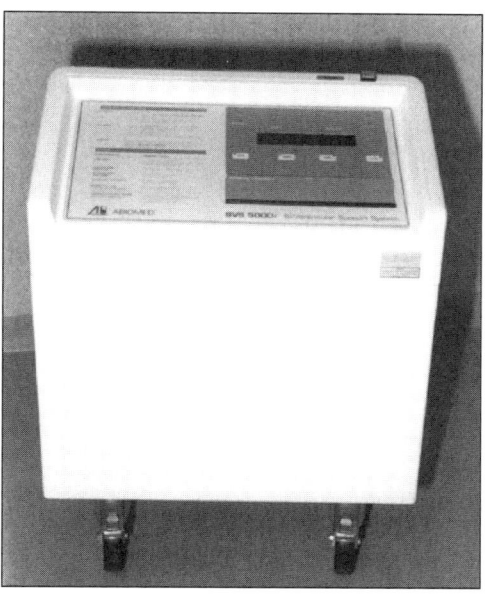

Fig. 2.3. The Abiomed BVS 5000 pneumatic drive console.

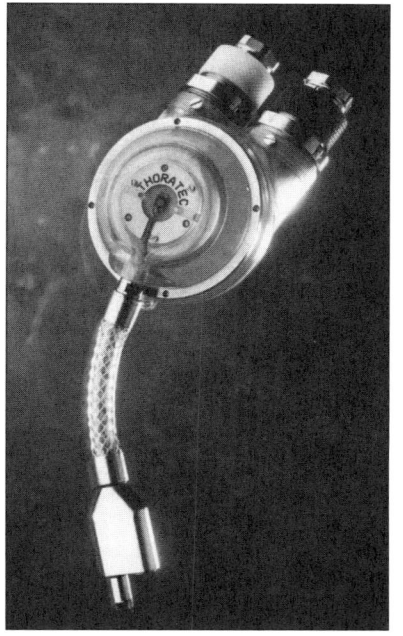

Fig. 2.4. The Thoratec ventricular assist device. Inlet and outlet cannulae are not shown. The pneumatic drive line exits the back of the blood pump. (Photograph courtesy of Thoratec Laboratories Corp.).

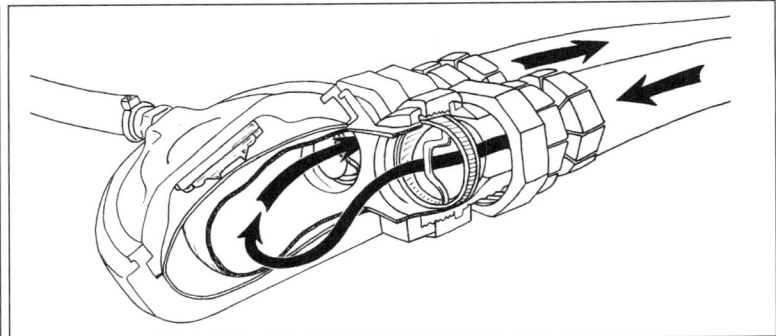

Fig. 2.5. Cross section of the Thoratec ventricular assist device. Tilting disc valves are positioned on the inlet and outlet sides of the blood pump. Intermittent pulses of air from the drive console periodically compress the blood sac to create the pumping action. (Illustration courtesy of St. Louis University Medical Center. Redrawn by Richard Manzer).

Fig. 2.6. The Thoratec ventricular assist device drive console.

Fig. 2.7. The Bio-Pump (Medtronic Bio-Medicus, Inc., Eden Prairie, MN) is prototypical of centrifugal pumps that have been used for mechanical ventricular assistance. (Photograph courtesy of Medtronic Bio-Medicus, Inc.).

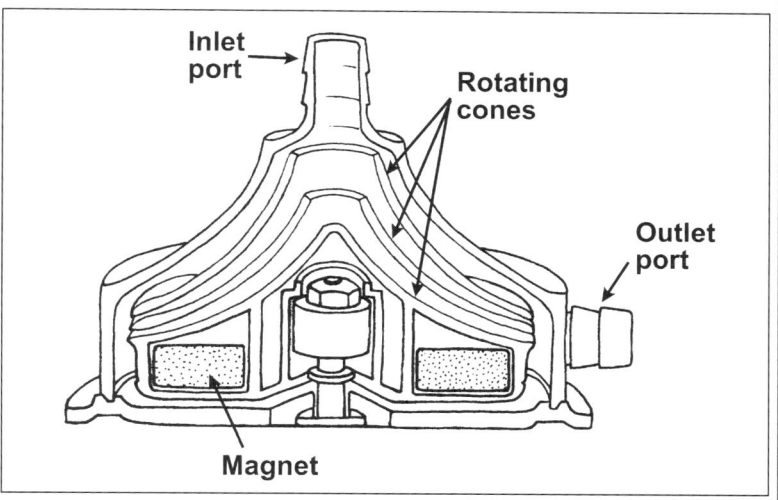

Fig. 2.8. The Bio-Pump centrifugal pump is constructed of a series of concentric cones. The cones are magnetically coupled to the drive console. The vortex created by the rapidly spinning cones moves blood from the inlet port at the apex of the blood pump to the outlet port at the base of the blood pump.

require custom designed cannulae the centrifugal pump can be connected to virtually any cannula employed for routine CPB. The centrifugal pumps are well known to most perfusionists and cardiac surgeons. These devices are not self-regulating and usually require personnel trained in the management of the drive unit in constant attendance at the patient's bedside (Fig. 2.9).

Fig. 2.9. The Bio-
Pump drive console.

SELECTING A VENTRICULAR ASSIST SYSTEM FOR USE AS A BRIDGE TO CARDIAC TRANSPLANTATION

Most cardiac transplant programs eventually decide to develop a mechanical blood pump program to allow potential transplant recipients who deteriorate hemodynamically to be supported with a device until a cardiac transplant can be performed.[8] These are frequently large urban centers with a well developed heart failure treatment program and personnel who are trained and designated to care for heart failure and heart transplant patients. Currently, 60-70% of patients undergoing cardiac transplantation are listed with the United Network for Organ Sharing as a Status 1. Status 1 designation is the highest acuity listing. Up to one-fourth of patients who undergo cardiac transplantation require an advanced form of mechanical circulatory support. As patient selection criteria have evolved and patient management techniques have improved there has been a dramatic increase in the number of centers that employ mechanical circulatory support as a bridge to cardiac transplantation. Using the above statistics the number of blood pumps required for this clinical application can be readily calculated for any given institution.

Two pneumatic, pulsatile VAD systems have been approved by the FDA for use as a bridge to cardiac transplantation (Table 2.4). The Thoratec VAD system is a paracorporeal device that can be configured for either right or left ventricular assistance.[9] Two blood pumps can be employed for biventricular assistance. When employed as a left VAD (LVAD), blood may be withdrawn from either the left

Table 2.4. Ventricular assist systems for use in patients requiring a bridge to cardiac transplantation

Device	Advantages	Disadvantages
Thoratec VAD System	Can be configured for right, left or biventricular support. Left atrial or left ventricular apex cannulation possible.	Patient must remain in the hospital. Requires systemic anticoagulation.
Thermo Cardiosystems, Inc. HeartMate 1000 IP LVAS	Only requires antiplatelet therapy.	Left heart support only. Patient must remain in the hospital.
Thermo Cardiosystems, Inc. HeartMate 1000 VE LVAS	Highly portable. Hospital discharge permitted. Only requires antiplatelet therapy.	Left heart support only.
Novacor N-100 LVAD	Highly portable. Hospital discharge permitted.	Left heart support only. Requires systemic anticoagulation.

atrium or left ventricular apex. Patient mobility is limited by the extracorporeal nature of the device and size of the external drive console. The Thermo Cardiosystems HeartMate 1000 IP LVAS (Thermo Cardiosystems, Inc., Woburn, MA) is an implantable, pneumatically powered, pulsatile VAD (Figs. 2.10, 2.11).[10] The device is tethered to the external drive unit by a percutaneous drive line (Fig. 2.12). Both systems are comparable in price. Although both VAD systems employ an external pneumatic drive console, the Thermo Cardiosystems HeartMate VAD allows greater patient mobility as the console can be maneuvered by the patient alone. The Thermo Cardiosystems device is only configured for left heart support and is designed for left ventricular apex cannulation alone. This consideration is not insignificant when caring for patients with biventricular dysfunction in which there is a potential need for right heart support following LVAD placement. This is also a consideration when caring for patients who have suffered an acute anterior wall myocardial infarction which potentially precludes left ventricular apex cannulation due to friable tissue or the presence of apical thrombus. A final consideration is that the paracorporeal position of the Thoratec VAD may allow implantation in a patient with a body surface area (BSA) less than 1.5 m². It is difficult, if not impossible, to insert an implantable VAD in a patient with a small body habitus (BSA < 1.5 m²).

In 1998, two electric, pulsatile VAD systems were approved by the FDA for use as a bridge to cardiac transplantation. Both the Thermo Cardiosystems HeartMate VE LVAS (Figs. 2.13, 2.14, 2.15)[10] and the Novacor N-100 LVAD (Novacor Division, Baxter Healthcare Corp., Santa Cruz, CA) (Figs. 2.16, 2.17, 2.18)[11] are implantable devices powered by a small external controller and battery pack. Both

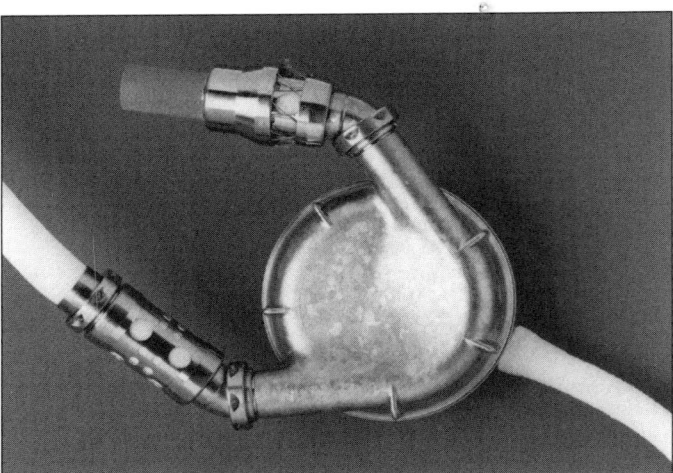

Fig. 2.10. The Thermo Cardiosystems 1000 IP (pneumatic) left ventricular assist device. The ventricular apex cannula is at the top of the photograph. The outlet graft is to the left. The drive line exits the lower right of the photograph. (Photograph courtesy of Thermo Cardiosystems, Inc.).

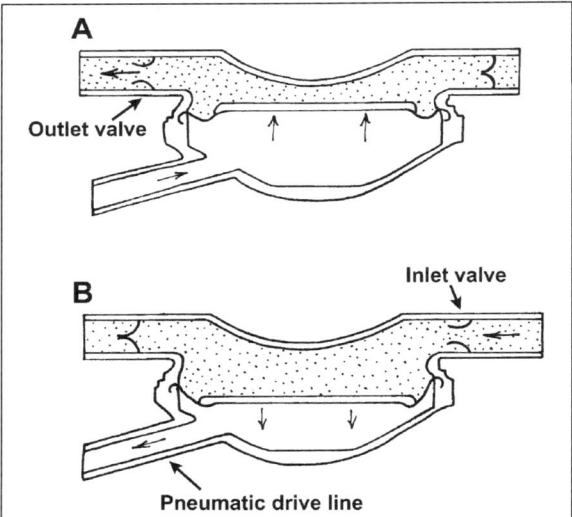

Fig. 2.11. Cross section of the Thermo Cardiosystems 1000 IP left ventricular assist device. A. During blood pump systole a pulse of air forces the pusher plate and diaphragm toward the rigid opposing surface. Blood is ejected through the outlet valve. B. During blood pump diastole, air is vented through the drive line. Blood enters the blood pump via the inlet valve.

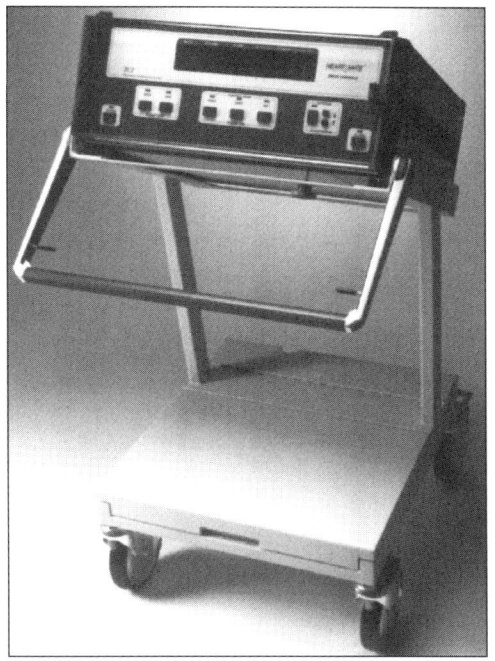

Fig. 2.12. The Thermo Cardiosystems 1000 IP LVAS pneumatic drive console. (Photograph courtesy of Thermo Cardiosystems, Inc.).

2

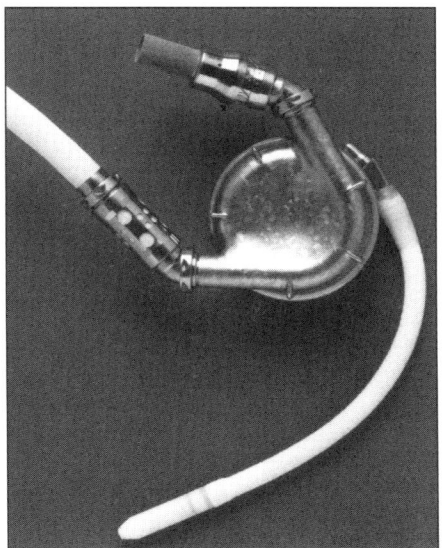

Fig. 2.13. The Thermo Cardiosystems VE (electric) left ventricular assist device. The ventricular apex cannula is at the top of the photograph. The outlet graft is to the left. The drive line is seen in the lower part of the photograph. (Photograph courtesy of Thermo Cardiosystems, Inc.).

2

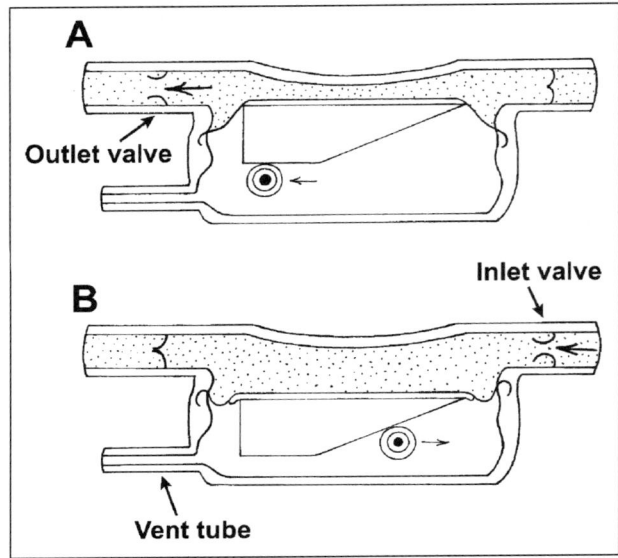

Fig. 2.14. Cross section of the implantable Thermo Cardiosystems VE left ventricular assist device. A. The integral electric motor drives the pusher plate and blood contacting diaphragm to create the pumping action. B. During blood pump diastole, blood enters the device as air exits the vent tube.

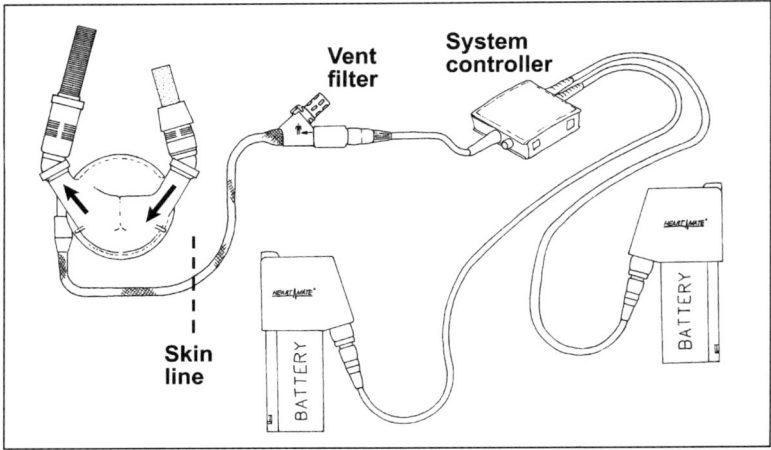

Fig. 2.15. The Thermo Cardiosystems VE LVAS. The operating system for this electric blood pump includes the controller that is worn on the patient's belt. The system is powered by two batteries that are worn in shoulder holsters.

2

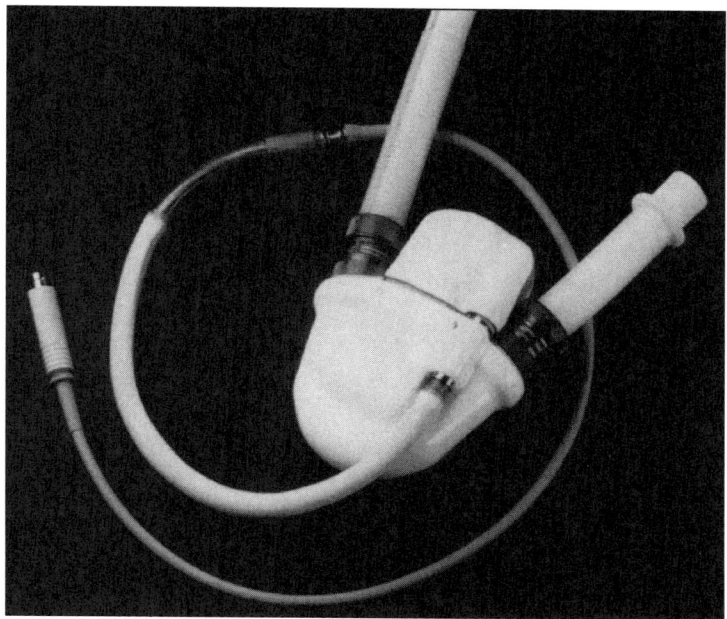

Fig. 2.16. The Novacor N-100 left ventricular assist device. (Photograph courtesy of the Novacor Division, Baxter Healthcare Corp.).

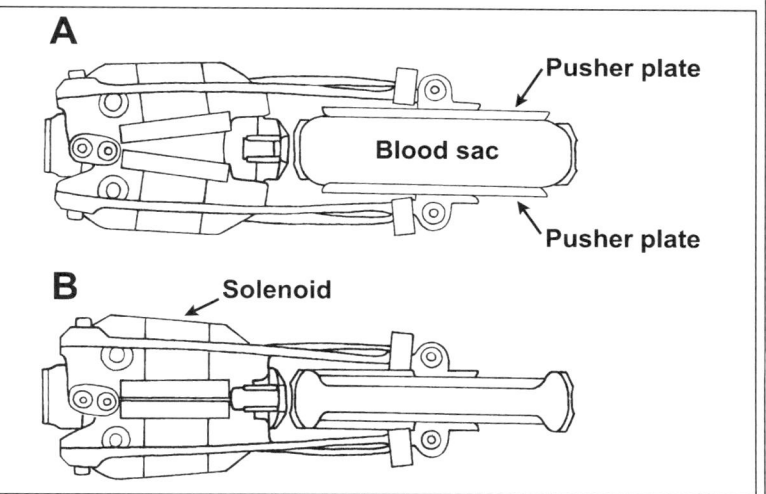

Fig. 2.17. Cross section of the Novacor N-100 left ventricular assist device. The electrome-chanical converter drives two pusher plates to periodically compress the blood sac. A. Blood pump diastole. B. Blood pump systole.

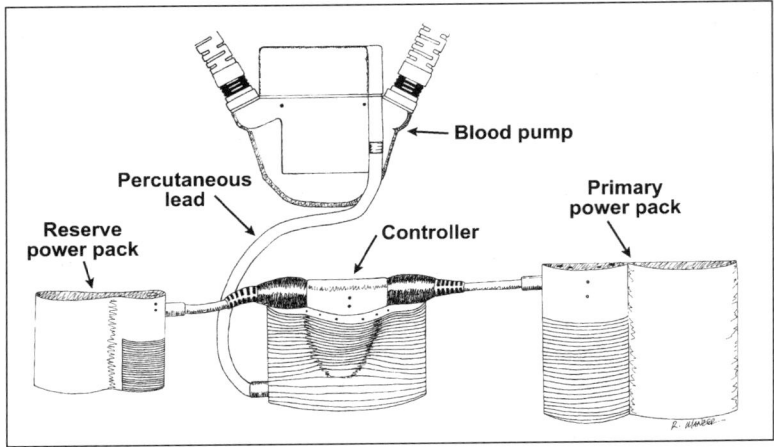

Fig. 2.18. The Novacor N-100 LVAD. The implantable blood pump is powered by two batteries and a system controller that are worn on a belt.

devices represent a dramatic improvement over the pneumatically powered VADs in that patients supported with an electric device are afforded a high degree of mobility and may be discharged from the hospital to await their cardiac transplant at home. Both devices are configured for left heart support utilizing a left ventricular apex cannula, an important consideration when caring for a patient with right heart failure or an anterior wall myocardial infarction as described above. Both the pneumatic and electric versions of the Thermo Cardiosystems HeartMate LVAD utilize the same blood pump design with an identical blood contacting surface. The surface is textured to promote a stable neointimal layer and patients are managed with antiplatelet therapy alone. Patients supported with either the Novacor or Thoratec VAD must be systemically anticoagulated.

Mechanical circulatory support as a bridge to cardiac transplantation is not necessarily unique to large urban transplant centers. A mechanical blood pump can be inserted in a postcardiotomy cardiogenic shock patient at a community hospital. If the patient does not recover ventricular function the patient can be transferred to a cardiac transplant center. In this instance the therapeutic endpoint changes from ventricular recovery to bridge to cardiac transplantation. If the patient has received the Thoratec VAD for postcardiotomy cardiogenic shock support the patient can simply be maintained on the same device until a transplant is performed. If, however, the patient has received a centrifugal blood pump or Abiomed device for postcardiotomy cardiogenic, shock then the patient will need to be converted to a long-term VAD system to improve patient mobility and enhance his rehabilitation potential while awaiting cardiac transplantation. The latter concept has recently been termed the "bridge-to-bridge" application. If a community hospital is developing a VAD program that allows them to care for postcardiotomy cardiogenic shock patients, early alliance with a regional transplant center greatly facilitates the care of patients who do not recover ventricular function. Preliminary contact with the transplant center prior to the purchase of a VAD system allows the clinicians at the community hospital to develop a program in conjunction with the mechanical blood pump program at the transplant center.

In general, all of the above considerations must be taken into account when selecting a VAD system. It is also important to take into account the expertise of the implanting surgeon, availability of support staff, expected utilization, financial impact upon the center and the population of patients that will benefit from the presence of such a program.

PHYSICAL PLANT CONSIDERATIONS

Once a VAD system is purchased arrangements must be made to provide maintenance and up-keep of the mechanical components. This necessary function can be accomplished in a variety of ways. The simplest, but perhaps the most expensive, is for the medical center to purchase a service contract with the VAD system manufacturer. Routine maintenance is scheduled regularly and when problems arise with a drive console or other system component the device manufacturer is contacted and service is provided by a company representative. Depending upon the VAD system manufacturer, a local biomedical engineering company may be hired to act on their behalf. Regardless, in-house expertise in device troubleshooting, repair and upgrade is not necessary. Alternatively, personnel from the medical center can be trained by the device manufacturer. Subsequently, all servicing is provided in-house. No service contract is required. At our institution we utilize the services of our Bioengineering Department which includes individuals who have Bachelors Degrees in Electrical Engineering and others who have two year electronics degrees. These engineers have been trained by the VAD manufacturer at hospital expense. Thereafter, all preventive maintenance and device repairs are performed on-site by hospital staff.

The blood pumps, drive consoles and associated components must be stored in an area that is readily accessible to operating room personnel. As this equipment is quite expensive it is also prudent to locate the storage area within the confines of the operating room where outside access is limited. Small components can be stored in a rolling cart and the expiration date on all equipment should be carefully monitored to ensure that a costly blood pump does not outdate.

Patients who are to undergo VAD implantation should be cared for in an operating room that is large enough to accommodate the patient, CPB machine, VAD associated equipment and drive unit, as well as extra personnel necessary for VAD implantation. These rooms must have quick-connect medical grade air and oxygen hook-ups for the CPB machine. Multiple, emergency power back-up outlets must be available in convenient locations in order to run the anesthesia equipment, CPB machine and VAD drive unit. Hemodynamic monitors should be positioned so that they may be easily viewed by all personnel involved in patient care. The monitors ideally have five channels available in order that the patient's electrocardiogram, systemic arterial pressure, left atrial pressure, central venous pressure and pulmonary artery pressures may be monitored simultaneously. The operating room should contain an antibacterial air filtration system and be under positive pressure with a minimum of 15 air changes per hour. Laminar air flow is ideal but not an absolute requirement.

The usual cardiac surgery operating room team is employed for VAD implantation. An extra perfusionist is required. One perfusionist is responsible for the conduct of CPB while a second perfusionist is responsible for the IABP console and VAD drive unit. Although the majority of CPB cases can be managed by two operating room nurses, one serving as a scrub nurse and one serving as a circulating nurse, VAD implantation usually requires an additional scrub nurse. Scrub nurse responsibility is divided. One nurse is located at the operating room table while the second scrub nurse is located at a back table for device preparation and assembly. VAD assembly usually requires preclotting of conduits, passivation of the pump surface with albumin or other component assembly. This is best performed at a separate sterile back table that is positioned in a protected area of the operating room away from the usual sterile operating room table.

Immediate postoperative care is provided in a large "isolation" room in the surgical intensive care unit. Although strict isolation is not necessary, access to the patient should be minimized in an effort to reduce traffic and hence infectious complications. The room must be large enough to accommodate the VAD drive console(s) and possibly an IABP console. The acuity of patient care in the immediate postoperative period can be quite high. Thus, 2:1 nursing is utilized for the first 24-48 hours following surgery. The perfusionist or other individual responsible for the VAD console is also asked to remain in close proximity to the intensive care unit. As the patient's condition stabilizes the patient is converted to 1:1 nursing with out-of-house coverage provided by the individual responsible for the VAD console.

When the patient has recovered from VAD insertion he is transferred to a private room on the general ward. Initially, the room is located near the nursing station to provide ready access to the patient and allow audible alarms on the VAD console to be easily monitored by nursing personnel. When the patient's rehabilitation has progressed and the patient or other family members have been instructed as to proper management of the VAD console, the patient is moved to a room that is located further away from the nursing station. Patients who receive mechanical blood pump support as a bridge to cardiac transplantation can spend inordinate amounts of time in the hospital. Thus, it is important to avoid extraneous distractions for these patients. By moving their room further away from the nursing station, the disturbing noise and disruptive activity associated with the routine care of other patients on the general ward can be eliminated. Privacy is of utmost importance. The room should be large enough to allow the patient and caregivers to move about easily despite the presence of the VAD console. Additional furniture should provide the patient with as home-like an atmosphere as possible. Accommodations should permit overnight family visitation.

TRAINING

When a VAD system is acquired by a hospital, principal individuals involved in the care of VAD patients are usually required to attend an off-site training program provided by the device manufacturer. The cost of this training program is usually built into the price of the VAD system. Travel and accommodations for the training session are provided by the hospital. In general it is helpful to have surgeons, perfusionists, nurses from the operating room, intensive care unit and general ward as well as transplant coordinators or cardiothoracic surgical clinical coordinators attend this session. When the VAD equipment arrives on-site a company bioengineer will check the equipment to ensure that it functions normally. Thereafter, additional on-site training is provided by the VAD manufacturer. The purpose of this additional training is to allow all individuals involved in the care of VAD patients to attend an abbreviated training program. Staff from each primary care area should be afforded the opportunity to attend this on-site training session. Coordinators for the mechanical blood pump program may subsequently provide ongoing inservicing to new personnel in the primary care areas and interval reviews for all associated hospital staff. In addition to the primary care area staff, clinical coordinators should train personnel from other care areas who may come in contact with VAD patients. These include but are not limited to cardiac rehabilitation and activity therapies staff as well as family members or other patient support individuals.

Periodic competency testing is mandatory. Routine inservicing should be scheduled on a 6 month basis with additional inservice updates as needed. The depth and frequency of the inservice program is dictated by the patient population, nursing turnover and needs expressed by hospital personnel. The competency testing consists of hands-on demonstration of the VAD console and associated hardware, problem solving with the VAD console as well as case studies and case management scenarios. Emergency protocols should be reviewed including alarm responses and who to contact should an emergency situation occur. Documentation of hospital staff participation in inservice programs and competency testing is maintained.

REFERENCES
1. Shinn JA. Nursing care of the patient on mechanical circulatory support. Ann Thorac Surg 1993; 55:288-294.
2. Moskowitz AJ, Weinberg AD, Oz MC et al. Quality of life with an implanted left ventricular assist device. Ann Thorac Surg 1997; 64:1764-1769.
3. Mehta SM, Aufiero TX, Pae WE et al. Results of mechanical ventricular assistance for the treatment of post cardiotomy cardiogenic shock. ASAIO J 1996; 42:211-218.

4.   Shook BJ. The Abiomed BVS 5000 biventricular support system. System description and clinical summary. Cardiac Surgery: State of the Art Reviews. 1993; 7:309-316.

5.   Jett, GK. Abiomed BVS 5000: Experience and potential advantages. Ann Thorac Surg 1996; 61:301-304.

6.   Holman WL, Bourge RC, McGiffin DC et al. Ventricular assist: Experience with a pulsatile heterotopic device. Semin Thorac Cardiov Surg 1994; 6:147-153.

7.   Noon GP, Ball JW, Papaconstantinou HT. Clinical experience with BioMedicus centrifugal ventricular support in 172 patients. Artif Organs 1995; 19:756-760.

8.   Hunt SA, Frazier OH, Myers TJ. Mechanical circulatory support and cardiac transplantation. Circulation 1998; 97:2079-2090.

9.   Farrar DJ, Hill JD. Univentricular and biventricular Thoratec VAD support as a bridge to transplantation. Ann Thorac Surg 1993; 55:276-282.

10.   Frazier OH, Myers TJ, Radovancevic B. The HeartMate left ventricular assist system. Overview and 12-year experience. Tex Heart Inst J 1998; 25:265-271.

11.   Vetter HO, Kaulbach HG, Schmitz C et al. Experience with the Novacor left ventricular assist system as a bridge to cardiac transplantation, including the new wearable system. J Thorac Cardiovasc Surg 1995; 109:74-80.

Intraaortic Balloon Counterpulsation

Wayne E. Richenbacher

INTRODUCTION

The concept of counterpulsation was introduced by Moulopoulos and colleagues in 1962 when they first described an intravascular counterpulsation balloon. Balloon counterpulsation was first employed clinically in 1968 by Kantrowitz et al. Intraaortic balloon pump (IABP) insertion originally required a femoral arteriotomy. In 1980, Bregman and Kaskel described percutaneous IABP insertion utilizing a sheath and dilators.[1] Unlike the ventricular assist device (VAD) and total artificial heart, the IABP is not designed to completely replace the function of the native ventricle. Rather, the IABP functions in concert with the native heart increasing coronary arterial perfusion while reducing myocardial oxygen consumption. At the same time, the IABP provides a modest increase in systemic perfusion. The IABP is a standard form of therapy for patients with a wide variety of cardiovascular diseases. In 1993, nearly 100,000 IABPs were inserted in the United States alone.

BALLOON PUMP DESIGN AND THEORY OF FUNCTION

The IABP is an intravascular, polyurethane membrane that is mounted on a catheter (Fig. 3.1). The balloon is available in sizes ranging from 30-50 ml; the standard size balloon is 40 ml. A central lumen allows passage of the balloon catheter over a small-diameter guidewire. The balloon is positioned within the descending thoracic aorta. A radiopaque marker on the end of the balloon facilitates radiographic identification of balloon location. The central lumen in the balloon

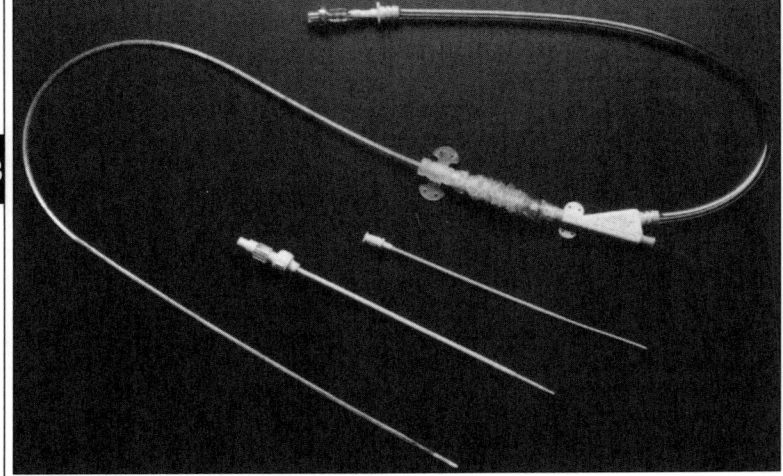

Fig. 3.1. The Datascope intraaortic balloon pump with associated protective sheath, dilators and console connector. (Photograph courtesy of Datascope Corp., Fairfield, NJ).

catheter allows subsequent monitoring of aortic blood pressure. The IABP is attached to a bedside console and triggered by the patient's electrocardiogram or arterial pressure curve (Fig. 3.2).

Intraaortic balloon counterpulsation serves three purposes (Fig. 3.3). Rapid, accurately timed balloon inflation and deflation results in an increase in coronary arterial perfusion, decrease in left ventricular afterload, and to a lesser extent, an increase in systemic perfusion. The IABP is pulsed in synchrony with the cardiac cycle. IABP inflation is timed to occur during ventricular diastole. As the balloon inflates after the aortic valve closes diastolic augmentation elevates the coronary perfusion pressure and increases coronary blood flow. Myocardial oxygen supply and the systemic perfusion pressure are increased. Balloon deflation occurs prior to ventricular contraction. The reduction in aortic end-diastolic pressure shortens the period of isovolumetric contraction and decreases left ventricular afterload. In addition to decreasing myocardial oxygen demand, a shortened period of isovolumetric contraction permits the aortic valve to open earlier allowing more time for ventricular ejection. The patient's stroke volume is increased and cardiac output enhanced. The net effect is an improvement in the ratio between myocardial oxygen supply and demand.

INDICATIONS FOR USE

The indications for IABP counterpulsation are listed in Table 3.1. Up to 6% of patients who undergo an open heart operation cannot be separated from

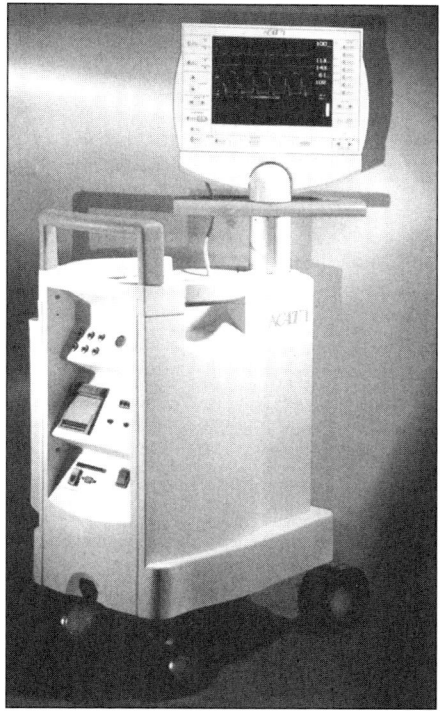

Fig. 3.2. The Arrow intraaortic balloon pump console. (Photograph courtesy of Arrow International, Reading, PA).

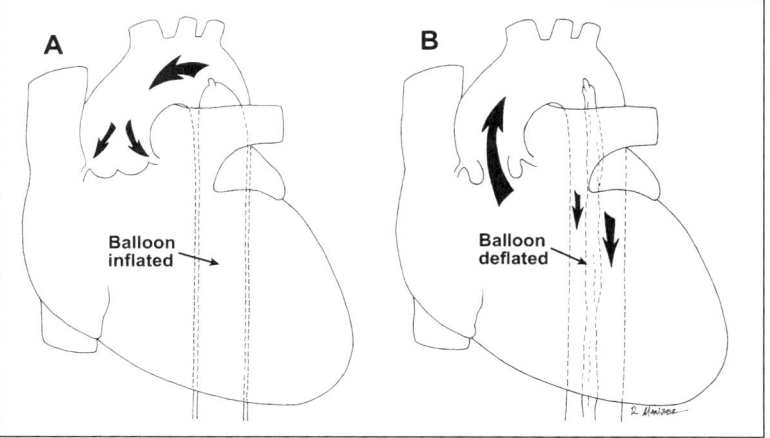

Fig. 3.3. Intraaortic balloon pump function. A. During cardiac diastole, balloon inflation increases coronary artery and systemic perfusion. B. Balloon deflation reduces left ventricular afterload during cardiac systole.

Table 3.1. Indications for intraaortic balloon counterpulsation

1. Refractory cardiogenic shock following cardiac surgery.
2. Refractory cardiogenic shock following an acute myocardial infarction.
3. Unstable angina.
4. Ischemia-related ventricular arrhythmias.
5. Mechanical problems within the heart.
6. Prophylactic support for high risk patients.

---

cardiopulmonary bypass (CPB) upon completion of the procedure despite inotropic support. Failing conventional medical management, IABP counterpulsation allows 60-80% of these patients to be weaned from CPB.[2-4] Cardiac function
generally improves within 1-2 days at which time the IABP can be removed.
Postcardiotomy cardiogenic shock that is refractory to medical therapy is usually
related to depressed preoperative left ventricular function, inadequate myocardial
preservation, a prolonged myocardial ischemic time, technical difficulties with
the conduct of the operation or an intraoperative or postoperative myocardial
infarction.

Approximately 15% of patients who suffer an acute myocardial infarction develop cardiogenic shock. In this clinical setting, cardiogenic shock may be related
to primary pump failure or mechanical complications of the acute myocardial
infarction. The latter include acute mitral regurgitation or a postinfarction ventricular septal defect. Acute mitral regurgitation may be due to papillary muscle
dysfunction or ruptured chordae tendineae. Three-quarters of patients who develop cardiogenic shock not amenable to conventional medical therapy following
an acute myocardial infarction improve hemodynamically following IABP insertion. Outcome is in large part determined by the underlying coronary artery pa-
thology. An early survival rate exceeding 90% can be achieved in patients with
operable coronary artery disease who undergo prompt revascularization. Patients
who develop a mechanical complication after an acute myocardial infarction are
best served by IABP counterpulsation followed immediately by cardiac catheterization and surgical repair. In patients with acute papillary muscle dysfunction
and mitral regurgitation the IABP reduces afterload, increases forward flow, stroke
volume and cardiac output. The resultant reduction in left ventricular preload
decreases pulmonary vascular congestion. IABP counterpulsation in patients with
a postinfarction ventricular septal defect decreases left ventricular afterload and
increases cardiac output while reducing the left-to-right shunt, right ventricular
pressure and central venous pressure.

Recently, it has been shown that IABP counterpulsation may be efficacious in
patients who do not fulfill the traditional hemodynamic selection criteria but who
suffer from unstable angina or ischemia related ventricular arrhythmias. It is
thought that the relief of angina following IABP counterpulsation in patients with
unstable angina may be related to an increase in coronary blood flow and reduction in myocardial oxygen demand. In general, patients with unstable angina receive IABP support in the face of deteriorating hemodynamics or ongoing ischemia

prior to myocardial revascularization. Ventricular tachyarrhythmias originating in the ischemic area surrounding an infarct zone may also respond to IABP counterpulsation. The reduction in ventricular tachyarrhythmias is thought to be due to an increase in myocardial perfusion and oxygenation in the ischemic zone.

Prophylactic IABP counterpulsation has been offered to "high risk" patients who are to undergo percutaneous transluminal coronary angioplasty (PTCA), cardiac catheterization, surgical revascularization or valve replacement as well as patients with ischemic coronary artery disease who are to undergo noncardiac surgical procedures. In patients who are to undergo an open heart operation it has been suggested that prophylactic placement of an IABP before anesthetic induction may be beneficial in patients who have a left ventricular ejection fraction below 25% or a left ventricular end-diastolic pressure greater than 20 mm Hg, a history of previous coronary revascularization, who are New York Heart Association Class III or IV or who have a left main coronary artery stenosis.[5] Preoperative prophylactic IABP placement is also thought to be beneficial in a patient with any of the above conditions who is scheduled to undergo a noncardiac operation. Theoretically, preoperative counterpulsation provides some degree of stability to the patient's hemodynamic status, permits a safer induction of general anesthesia and controls myocardial ischemia prior to surgical revascularization. At our institution, a prophylactic IABP is employed in the cardiac catheterization laboratory when a patient is to undergo PTCA of a single remaining coronary artery or a complex coronary arterial stenosis.[6] The latter would include irregular, calcified plaques or bifurcation lesions in which the potential for intimal injury or flow interruption is increased. A prophylactic IABP is occasionally inserted prior to cardiac catheterization in a patient with major vessel disease associated with severe left ventricular dysfunction. An alternative management scheme in the patient who is to undergo a high risk catheter-based intervention is to cannulate the contralateral femoral artery with a small-diameter catheter. Should the patient develop hemodynamic instability during the cardiac catheterization or PTCA an IABP may be quickly inserted into the opposite femoral artery. Prophylactic preoperative IABPs are rarely used at our institution and should never serve as a substitute for a careful cardiac anesthetic.

CONTRAINDICATIONS FOR USE

Contraindications to the use of IABP counterpulsation are listed in Table 3.2. As IABP inflation occurs during diastole a patient with aortic insufficiency would

Table 3.2. Contraindications to intraaortic balloon counterpulsation

1. Aortic insufficiency.
2. Aortic dissection.
3. Abdominal aortic aneurysm.
4. Aortoiliac or femoral arterial occlusive disease.
5. Percutaneous insertion in a patient who has had a recent groin incision.

experience a retrograde flow of blood across the regurgitant aortic valve into the left ventricle during balloon inflation. Left ventricular work and myocardial oxygen consumption would increase as would left ventricular preload. In this instance cardiac output would fall. IABP counterpulsation should not be employed in a patient with an aortic dissection as the balloon may be inserted into the false lumen exacerbating the aortic injury. Balloon counterpulsation should not be employed in a patient with an abdominal aortic aneurysm for fear the balloon may precipitate rupture of the aneurysm. Similarly, an IABP should not be inserted in a patient with severe calcific aortoiliac or femoral arterial occlusive disease. It may be difficult to advance a balloon in a patient with peripheral vascular disease and the presence of the balloon in the diseased vessel may precipitate intravascular thrombosis or distal embolization with limb ischemia. Intimal calcification in the aorta or distal arterial tree may damage the balloon during insertion or counterpulsation. The percutaneous insertion technique should not be used in a patient who has had a recent groin incision with violation of the soft tissues at the proposed puncture site. Such a patient is at significant risk for bleeding into the soft tissues.

INSERTION TECHNIQUE

PERCUTANEOUS INSERTION TECHNIQUE

The IABP is most commonly inserted in a percutaneous fashion via the common femoral artery. Pre-insertion evaluation of the patient's femoral arterial and pedal pulses ensures that the IABP is inserted in the leg with the best arterial inflow and facilitates rapid recognition of limb ischemia following balloon insertion. Strict aseptic technique is employed during IABP insertion. The common femoral artery is accessed using the Seldinger technique (Fig. 3.4). An 18 gauge needle is inserted into the common femoral artery and a flexible J-tipped guidewire passed through the needle and advanced into the descending thoracic aorta. The femoral arterial puncture should occur below the inguinal ligament to avoid a transperitoneal puncture and above the profunda femoris artery to reduce the potential for leg ischemia related to superficial femoral arterial cannulation. The access needle is withdrawn and the skin adjacent to the guidewire incised with a #11 scalpel. A small caliber dilator is advanced over the guidewire. The initial dilator is removed and a larger dilator-sheath assembly advanced over the guidewire. The larger dilator is withdrawn and the IABP advanced over the guidewire through the sheath.

The IABP comes prewrapped in a tightly folded configuration. The fold system was developed to minimize the balloon profile. Prior to removing the IABP from the packaging system a vacuum is applied to the balloon to remove all atmospheric air thereby reducing the diameter of the balloon to its smallest dimension. Vacuum is applied to the IABP by attaching a 50 ml syringe and a one way valve to the balloon catheter and withdrawing the plunger. The syringe is removed leaving the one way valve attached to the IABP. Prior to inserting the IABP through the sheath the length of IABP to be inserted is estimated by marking the shaft at the insertion site while holding the tip of the IABP level with the patient's sternal angle. The guidewire is passed through the central lumen of the IABP and the IABP advanced through the sheath and positioned in the descending thoracic aorta (Fig. 3.5). To provide maximal therapeutic benefit the tip of the IABP should

3

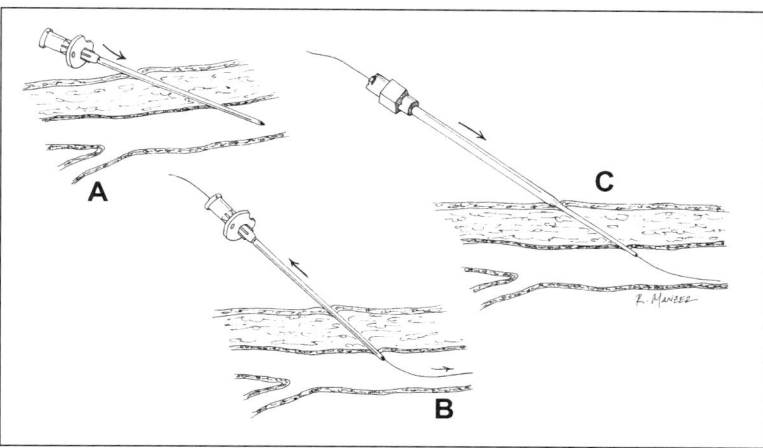

Fig. 3.4. The Seldinger technique for vascular access. A. The vessel is accessed percutaneously with a small-diameter needle. B. A guidewire is inserted through the needle into the vessel. The needle is withdrawn. C. A large-diameter dilator is advanced over the guidewire into the vessel.

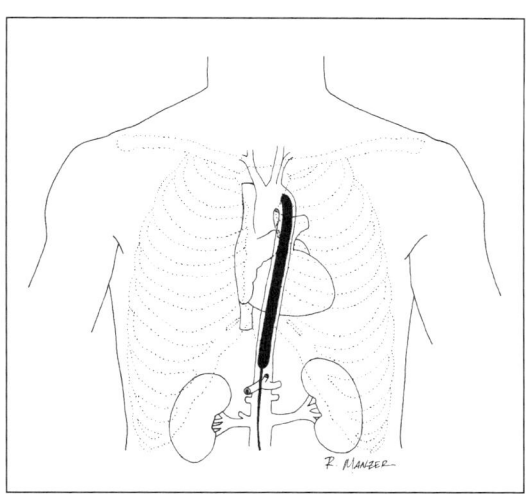

Fig. 3.5. The intraaortic balloon is positioned in the descending thoracic aorta. The tip of the balloon catheter is 2 cm distal to the origin of the left subclavian artery. The proximal end of the balloon is positioned above the origin of the renal arteries.

be located 2 cm distal to the origin of the left subclavian artery. The distal end of the IABP should be located above the origin of the renal arteries. If the IABP is inserted in the cardiac catheterization laboratory, fluoroscopy is used to verify satisfactory balloon positioning. If the IABP is inserted in the operating room or intensive care unit the balloon position is determined by a chest roentgenogram upon completion of the insertion procedure. The presence of a sterile protective sheath over the IABP catheter superficial to the insertion site allows minor adjustments in balloon position without violating the sterility of the balloon shaft adjacent to the insertion site (Fig. 3.6). The balloon is anchored to the patient's leg with silk sutures at both the entry point and at the distal connector. The balloon is unwound, purged, connected to the bedside console and pulsed.

SHEATHLESS INSERTION TECHNIQUE

In the nonanticoagulated patient a "sheathless" insertion technique may be utilized. Hemostasis at the puncture site is not as good with sheathless insertion as it is with insertion using a sheath. However, the sheathless insertion technique minimizes the obstruction to flow in the common femoral artery and is particularly applicable in diminutive patients, patients with known or suspected peripheral vascular disease or women in whom the arteries are disproportionately smaller when compared to a man of similar weight. Sheathless IABP insertion utilizes the same Seldinger technique to gain femoral arterial access as described above. The initial small-diameter dilator is inserted over the guidewire. The small-diameter

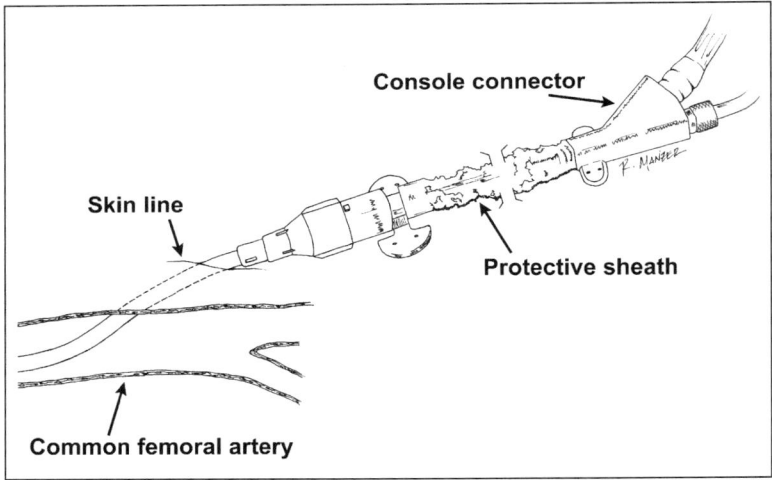

Fig. 3.6. The intraaortic balloon catheter enters the common femoral artery proximal to the bifurcation into the superficial and deep branches. Several centimeters of the balloon catheter adjacent to the insertion site are housed in a protective sheath. The protective sheath allows minor adjustments in the balloon position without violating the sterility of the balloon catheter.

dilator is withdrawn and the IABP catheter advanced over the guidewire. By eliminating the second large dilator and sheath assembly, the diameter of the arterial puncture is decreased and only the balloon catheter remains in the common femoral artery. The potential for limb ischemia and bleeding at the time of IABP removal are decreased. Most recently, low profile IABPs using a sheath as small as 8 F in diameter have been developed. The central catheter in the low profile balloon is more flexible and smaller in diameter allowing these balloons to be inserted with a single dilator and sheath assembly. Of note, the low profile balloons require a smaller guidewire than that which is employed with conventional balloons that use the two-dilator insertion technique.

Open Insertion Technique

If the common femoral artery cannot be accessed via a percutaneous technique the IABP may be inserted into the vessel using an open surgical technique. The common femoral artery is exposed and a longitudinal arteriotomy created. A 5 cm segment of an 8-10 mm diameter vascular graft is anastomosed in an end-to-side fashion to the common femoral artery (Fig. 3.7). The vascular graft is oriented at a 45° angle with respect to the femoral artery. The IABP is passed through the vascular graft into the artery and advanced into the descending thoracic aorta. The IABP is fixed in position by tying two separate umbilical tapes around the vascular graft. The central catheter is brought through the most caudad aspect of the skin incision and the wound closed in layers. The vascular graft should be located deep to the skin closure.

Transthoracic Insertion Technique

If an abdominal aortic aneurysm or peripheral vascular disease precludes femoral arterial insertion, the IABP may be inserted directly into the ascending aorta

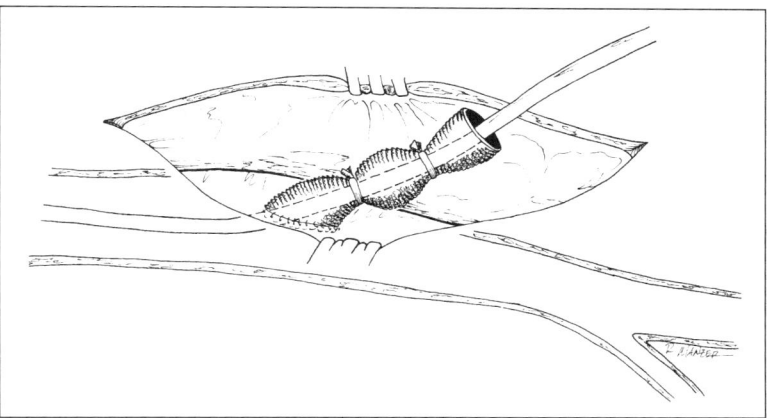

Fig. 3.7. The open technique for intraaortic balloon pump insertion. The balloon catheter is advanced through a vascular graft that is anastomosed to the common femoral artery.

or transverse aortic arch (Fig. 3.8).[7] Access to these insertion sites is through a median sternotomy usually at the time of the patient's open heart operation. A partial occlusion clamp is applied to the anterior or right anterolateral aspect of the ascending aorta. A small, longitudinal aortotomy is created, and a vascular graft is anastomosed to the entry site in a manner identical to that described in the open femoral arterial technique. The balloon is inserted into the vascular graft, the ascending aorta or transverse arch and advanced into the descending thoracic aorta. In order to facilitate IABP advancement the left pleural space is opened and digital pressure applied to the origin of the left subclavian artery. The IABP can be manually directed into the descending thoracic aorta. Alternatively, a pledgeted 3-0 Prolene suture (Ethicon Inc., Somerville, NJ) is placed in a horizontal mattress fashion in the aorta. A #11 scalpel is used to puncture the aorta and the balloon catheter is inserted directly into the ascending aorta and advanced into the descending thoracic aorta. The mattress suture is snared and the snare tied to the balloon catheter. Regardless of technique employed, the IABP exits the chest through the caudad aspect of the sternotomy incision. If bypass conduits are compressed by the IABP catheter, the IABP may exit the chest via the right second intercostal space. The exit site location must be determined prior to inserting the IABP into the aorta. If the catheter is to exit the chest via the right second intercostal space, the IABP must be passed through the chest wall prior to inserting the

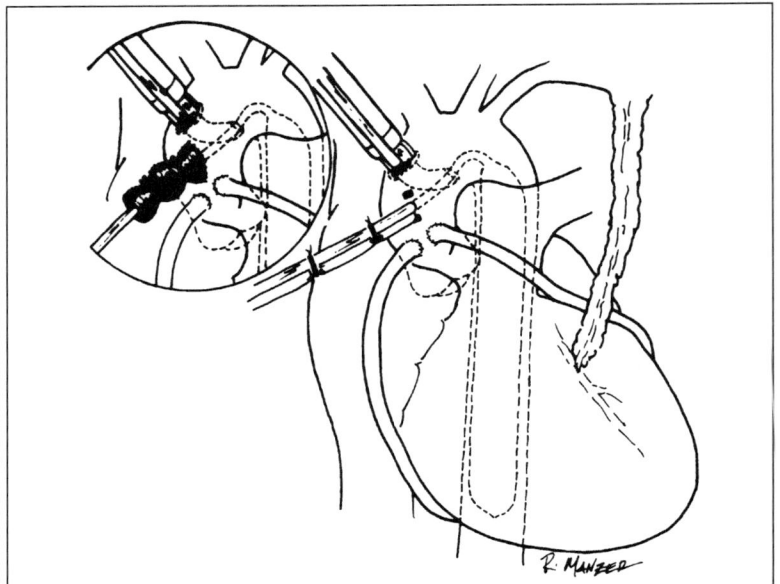

Fig. 3.8. Transthoracic intraaortic balloon pump insertion. The balloon catheter can be inserted through a pursestring suture or advanced through a small-diameter vascular graft that has been anastomosed to the ascending aorta (inset).

catheter into the ascending aorta. As the IABP is inserted in an antegrade fashion versus retrograde advancement with femoral arterial insertion, the proximal end of the balloon should be located distal to the left subclavian artery. As there is no radiopaque marker at this location, satisfactory balloon positioning is determined with the aide of transesophageal echocardiography.

OTHER INSERTION SITES
In patients in whom long-term support is anticipated, the IABP may be inserted into the subclavian or iliac artery. These insertion sites are mentioned only out of historical interest as development of the VAD has precluded the need for long-term IABP support.

MANAGEMENT OF THE INTRAAORTIC BALLOON PUMP PATIENT

PATIENT CARE
Following IABP insertion the insertion site is covered with an occlusive Tegaderm (3M Health Care, St. Paul, MN) dressing. The sterile dressing is changed daily. The dressing change includes swabbing the insertion site and surrounding skin with a povidone-iodine solution. Povidone-iodine ointment is applied to the entry point and the sterile occlusive dressing reapplied.

If the common femoral artery has been employed as the insertion site, the patient must remain in a supine position and the head of the bed elevated no more than 30°. If the patient is noncompliant or agitated the ipsilateral knee is immobilized with a soft splint. Both maneuvers reduce the likelihood of an intimal injury to the femoral artery at the point of insertion.

Pedal pulses in the ipsilateral leg are assessed every 2 hours. The balloon shaft is examined for blood, diagnostic of balloon rupture, every 2 hours. The frequency of pulse checks is increased to every hour if the patient has a history of peripheral vascular disease. The anticoagulation regimen varies with the indication for insertion. Cardiotomy patients who receive an IABP in the operating room or in the immediate postoperative period usually do not require anticoagulation. Even if the systemic anticoagulation employed during CPB has been reversed, the patient will still have the post-pump coagulopathy related to platelet dysfunction. Following IABP insertion the central lumen used for monitoring the aortic blood pressure is flushed in a manner identical to any arterial pressure line. A continuous heparin sodium infusion (1000 U heparin sodium in 500 ml normal saline) is administered at a rate of 3 ml/hr. If the IABP is inserted in the cardiac catheterization laboratory in a noncardiotomy patient 5000 U of heparin sodium are administered intravenously at the time of balloon insertion. As the potential for hemorrhage at the insertion site is reduced in a patient who has not been on CPB the subsequent central lumen heparin sodium flush is modified. Heparin sodium (5000 U in 500 ml normal saline) is administered through the central lumen at 3 ml/hr.

COUNTERPULSATION TIMING

The most critical component of IABP patient management is proper timing of balloon counterpulsation. The balloon pulse is synchronized with the cardiac cycle. The ratio between myocardial oxygen supply and demand is maximized when balloon inflation occurs during ventricular diastole and balloon deflation occurs prior to ventricular contraction. In order to properly time IABP pulsation it is important to understand the relationship between the electrical and mechanical activity of the heart as demonstrated in the electrocardiogram and associated arterial pressure trace (Fig. 3.9). The QRS complex on the electrocardiogram signifies ventricular depolarization. Ventricular depolarization initiates the mechanical event, ventricular contraction. Initially, ventricular contraction occurs with the aortic valve closeS. This period of isovolumetric contraction takes place as the ventricular pressure rises. Isovolumetric contraction ceases when the ventricular pressure exceeds the aortic pressure and the aortic valve opens. Thereafter, ventricular ejection occurs and there is a rise in the arterial pressure curve known as the anacrotic limb. The peak systolic pressure is the highest point on the arterial pressure curve. Following ventricular ejection the ventricle begins to relax and the arterial pressure falls. When the pressure within the left ventricle falls below the aortic pressure the aortic valve closes. The point at which the aortic valve closes is marked on the arterial pressure trace by the dicrotic notch. The dicrotic notch occurs as a result of the elastic recoil of the aortic wall following aortic valve clo-

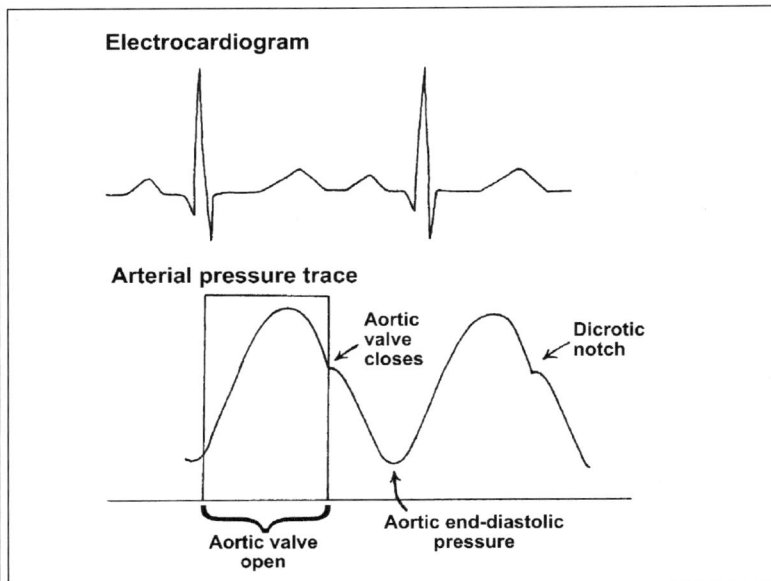

Fig. 3.9. The relationship between the electrocardiogram (top) and arterial pressure trace (bottom) in a patient who is NOT receiving intraaortic balloon pump support.

sure. As arterial runoff concludes the aortic pressure continues to fall. The lowest point on the arterial pressure trace signifies the aortic end-diastolic pressure.

In order to properly time the IABP, the peak systolic pressure, dicrotic notch and aortic end-diastolic pressure must be located on the arterial pressure curve. Balloon inflation should occur during ventricular diastole and the timing for balloon inflation should occur immediately following aortic valve closure (Fig. 3.10). Thus, the IABP console should be set to inflate the balloon at the dicrotic notch on the arterial pressure trace. Balloon deflation should occur before the start of the subsequent ventricular systole. The period of optimal balloon deflation should be timed to occur at the aortic end-diastolic pressure on the arterial pressure trace. In brief, the balloon should be timed to inflate between the closing and opening of the aortic valve when it will not compete with ventricular ejection. The balloon should **not** be inflated at any time during ventricular systole. The mechanical effect of balloon inflation and deflation is readily seen in the arterial pressure trace (Fig. 3.11). Balloon inflation results in a rise in arterial pressure following the dicrotic notch. This increase in aortic pressure occurs during ventricular diastole. Deflation of the balloon prior to ventricular contraction results in a brisk decline in aortic pressure.

Timing of IABP counterpulsation is most readily accomplished with the IABP synchronized 1:2 with the patient's arterial pressure trace. Once the appropriate inflation and deflation times are determined, augmentation is set at 1:1. The timing process is simplified if the arterial line is located in the aortic root as the temporal relationship between aortic valve closure and the dicrotic notch on the pressure trace is correct. As the pressure wave form takes time to travel down the arterial tree, IABP inflation and deflation must be set earlier in the arterial trace when this trace is taken from a site distant from the aortic root. Proper timing ensures that the IABP provides the greatest therapeutic benefit. Early inflation results in premature closure of the aortic valve. Incomplete ventricular emptying may lead

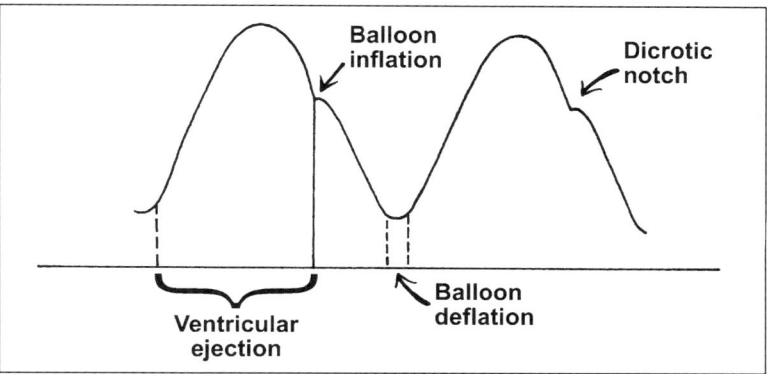

Fig. 3.10. The arterial pressure trace demonstrating the points at which intraaortic balloon inflation and deflation should occur. The balloon should never be inflated during ventricular ejection.

Fig. 3.11. Proper intraaortic balloon counterpulsation results in an elevation in the peak systolic pressure and fall in the aortic end-diastolic pressure.

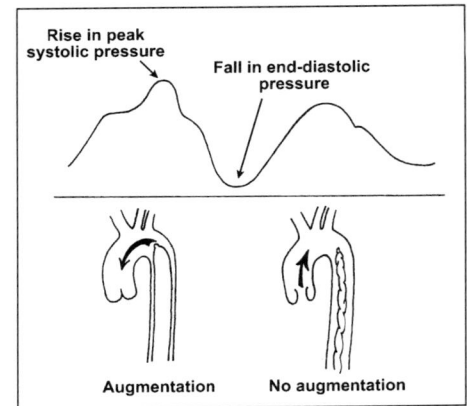

to elevated preload while the reduction in stroke volume will decrease cardiac output. Late inflation is not as hazardous as early inflation. However, late inflation will not allow the IABP to offer maximal hemodynamic support. Late inflation decreases the coronary artery perfusion pressure and results in a lower diastolic augmentation pressure. If the IABP deflates early there is no reduction in afterload and, hence, no decrease in myocardial oxygen consumption. Late IABP deflation is particularly counterproductive in that afterload is increased along with myocardial oxygen consumption. There is a reduction in systolic ejection and cardiac output.

Timing is initially set following balloon insertion. Timing is subsequently checked every 8 hours. If the initial timing is determined while the patient is either tachycardic or bradycardic, no timing changes are necessary provided the patient's heart rate remains within 20% of the original heart rate. In practice, the IABP is originally timed using both the electrocardiogram and arterial pressure trace. Thereafter, the most frequent trigger source for the IABP is the electrocardiogram. Use of the electrocardiogram as the trigger source results in fewer alarm conditions from the IABP console. If the arterial pressure trace is used as the trigger source an alarm condition occurs each time the pressure trace is lost, as during blood sampling. Most balloon consoles incorporate electrosurgical interference suppression (ESIS) which should theoretically allow the balloon to function in the electrocardiogram mode during the use of electrocautery. In practice, ESIS is not particularly effective, and it is simpler to change the trigger source to the arterial pressure trace when electrocautery is employed.

Should the patient lose arterial blood pressure or cardiac rhythm as in an arrest situation the IABP console can be placed on an internal trigger. In general, the internal trigger provides a fixed rate of balloon inflation with the rate varying between 40 and 120 pulses per minute. The internal trigger can also be utilized during CPB to provide some degree of pulsatility to the systemic blood flow. The balloon console should never be placed in the internal trigger mode if the patient is generating a cardiac output. Should a patient require defibrillation, this procedure can be performed without disconnecting the balloon console as the system is electrically isolated from the patient.

TRANSPORTATION OF THE PATIENT REQUIRING INTRAAORTIC BALLOON PUMP SUPPORT

3

Patients who require IABP counterpulsation must frequently be moved from one area of the hospital to another or from one hospital to another. In-hospital transportation is necessary when a patient must be moved between the cardiac catheterization laboratory, operating room, intensive care unit or other diagnostic area. Equipment required for the care of the patient should be consolidated and whenever possible attached to the patient's bed. The transfer should occur via the shortest route through the hospital utilizing the least congested corridors. A designated individual should precede the patient to hold the elevator for transfer, and it is important to ensure that the elevator is large enough to accommodate the bed, balloon console and transfer personnel. All preparations for transfer should be completed prior to converting the balloon console to the battery mode. Check the manufacturer's information to determine the battery mode time for the console. In general, the balloon console can function in the battery mode for 90 minutes. Upon arrival at the destination area the IABP console should be immediately connected to a wall outlet in order to preserve the console battery life. At least two individuals should be available to transfer the patient within the hospital. The primary nurse is responsible for the patient's care. The primary responsibility of a perfusionist or nursing assistant is to ensure that the IABP functions appropriately during transport. If the patient is intubated or the acuity of care exceeds the capabilities of two individuals, a third healthcare provider should be available for transport. In general, a respiratory therapist is included.

Interhospital transfer requires a correspondingly greater degree of preparation. Ground transportation is accomplished in an ambulance that must be large enough to accommodate the patient, bed, transport personnel and IABP console. The transport personnel must have ready access to both the patient and IABP console. An alternating current (AC) power source must be available in the ambulance. Most ambulances utilize an invertor to convert direct current power generated by the ambulance to AC power. The invertor can be employed to power the IABP console only if the power output from the invertor exceeds the power requirement of the IABP console. If an invertor is not available the transport time cannot exceed the operational time provided by the console battery.

When the interhospital transfer is to be accomplished by air the patient may travel in either a fixed wing or rotor wing aircraft.[8] Again, the aircraft configuration must accommodate the patient, equipment and transport personnel. The power requirement is similar to that necessary for ambulance transportation. An invertor must be able to provide the AC power required by all equipment utilized in the transfer. Air transportation creates a number of unique situations both with balloon console function and patient management. Low frequency vibrations in rotor winged aircraft can interfere with the electrocardiogram signal. Thus, during aircraft transportation the balloon should be triggered from the arterial pressure trace. Fixed wing aircraft acceleration and deceleration during take-off and landing, respectively, can affect patient hemodynamics depending upon the orientation of the patient with respect to the aircraft. If the patient's head is oriented toward the front of the aircraft, a positive acceleration position, a fall in both cardiac output and stroke volume can occur during take-off due to a reduction in preload. If on the other hand, the patient is oriented with the feet toward the front of the aircraft, a negative acceleration position, preload may increase during take-off as more blood is forced toward the heart. This increase in preload can precipitate acute cardiac failure. These changes in the clinical condition of the patient can be minimized by requesting that the pilot keep acceleration and deceleration forces to a minimum while avoiding steep flight angles. Shifts in intravascular volume can be temporarily treated with pharmacologic support during take-off and landing.

Although the cabin is pressurized in a fixed wing aircraft, changes in atmospheric pressure during take-off and landing can result in variations in the degree to which the balloon fills. According to Boyle's Law, if the temperature remains constant the volume of gas varies inversely to the pressure acting upon it. As pressure decreases, gas expands and vice versa. As the aircraft ascends the barometric pressure falls. The gradient between the inside and outside of the IABP will increase and the balloon will expand (Table 3.3). Although most balloon consoles will monitor and self-adjust for changes in the volume of gas sent to the balloon,

---

Table 3.3. Theoretical estimates of changes in balloon volume with increased altitudes calculated from the Table of US Standard Atmosphere

$$\frac{P_1V_1 + 80 \text{ mm Hg MAP}}{\text{Temperature}} = \frac{P_2V_2 + 80 \text{ mm Hg MAP}}{\text{Temperature}}$$

Altitude (FT)	Volume of Unrestrained Gas (ml)	Barometric Pressure (mm Hg)
0	40	760
2000	43	706
4000	46	656
6000	50	609
8000	54	565
10000	58	523
12000	63	484
14000	68	446

P = Pressure, V = Volume, MAP = mean arterial pressure
(Reproduced with permission from: Mertlich G, Quaal SJ. Crit Care Nurs Clin North Am 1989; 1:443-457)

in order to avoid balloon rupture it is prudent to purge and refill the balloon every 2,000 feet during ascent and every 1,000 feet during descent.

WEANING FROM INTRAAORTIC BALLOON PUMP SUPPORT

Although intraaortic balloon counterpulsation provides obvious therapeutic **3** benefit for the patient with hemodynamic instability or unstable angina the IABP is also associated with a number of potential complications. Thus, as soon as a patient achieves some semblance of stability it is wise to wean and terminate IABP support. Following IABP insertion, the balloon console is set at a 1:1 assist ratio; the IABP augments every beat of the patient's cardiac cycle. There are two schools of thought with respect to timing of IABP weaning. One approach is to leave the patient on a modest amount of inotropic support with plans to remove the IABP as early as possible. A second approach is to take time to wean the patient from inotropic support with plans to remove the IABP when inotrope administration has been decreased to a minimum level or discontinued altogether. We prefer the latter approach. Should the patient deteriorate following IABP removal it is possible to add a low dose inotrope to the patient's medical regimen to again achieve hemodynamic stability. Furthermore, in the patient with primary ventricular dysfunction the presence of the IABP affords the patient the opportunity to recover ventricular function prior to balloon removal. Obviously, should the patient develop a complication of IABP counterpulsation the balloon is removed when the complication occurs.

To wean from IABP support, balloon augmentation is serially decreased. The assist ratio is reduced to 1:2 during which time the IABP augments every other cardiac cycle. This level of support is maintained for several hours as the patient is observed for signs of hemodynamic decompensation. The signs include a fall in systemic blood pressure, rise in left-sided filling pressures, fall in cardiac and urine output, the development of angina, ischemic changes in the electrocardiogram or ventricular arrhythmias. If the patient tolerates the decrease in augmentation ratio IABP, counterpulsation is further reduced to 1:3 or 1:4. The IABP is removed when the patient is able to maintain an adequate cardiac output index (> 2.0 L/min/m^2) and reasonable left-sided filling pressures (left atrial or pulmonary capillary wedge pressure < 10-15 mm Hg) on minimal inotropic support (Table 3.4).

INTRAAORTIC BALLOON PUMP REMOVAL

If systemic anticoagulation was employed during the period of IABP support, the anticoagulant is discontinued and coagulation studies allowed to normalize prior to balloon removal. The insertion site is prepped and draped in a sterile fashion. The individual withdrawing the balloon should wear a gown, gloves and protective eyewear due to the potential for blood exposure. A designated assistant

---

Table 3.4. Criteria to be met before intraaortic balloon pump removal

1. Counterpulsation assist ratio 1:3 or 1:4.
2. Minimal or no inotropic support.
3. Cardiac output index > 2.0 L/min/m^2.
4. Systemic arterial blood pressure > 100 mm Hg.
5. Left atrial or pulmonary capillary wedge pressure < 10-15 mm Hg.
6. Urine output > 30 ml/hr.
7. No angina.
8. No ischemic changes on electrocardiogram.
9. No new ventricular arrhythmias.

---

converts the IABP console to the stand-by mode. This individual removes the drive line from the balloon console and aspirates any remaining gas from the balloon using a 50 ml syringe. Aspiration of the gas from the IABP reduces the diameter of the balloon to a minimum facilitating withdrawal.

The balloon catheter is withdrawn until the balloon touches the introducer. No attempt should be made to pull the balloon itself into the introducer sheath. Manual pressure is applied to the femoral artery distal to the insertion site. The balloon and sheath are withdrawn as a single unit and blood is permitted to eject from the insertion site for one or two heartbeats. This allows any thrombotic debris to be ejected from the vascular space. Manual pressure is then applied to the insertion site for 30 minutes. Thereafter, pressure is applied to the insertion site for 8 hours using a 5-10 pound sandbag. To ensure that the patient does not move the ipsilateral leg during this time a soft knee immobilizer may be applied. It is important to check pedal pulses during the period of time that pressure is applied to the groin following balloon removal. Pressure applied to the femoral artery should achieve hemostatis but not jeopardize limb perfusion.

If the IABP has been inserted using the open technique, withdrawal requires surgical exploration with balloon and vascular graft removal (Table 3.5). In the operating room using sterile technique the groin incision is reopened. The common femoral artery is clamped distally and the umbilical tapes on the vascular graft removed. The balloon is disconnected from the console, aspirated and withdrawn. A second vascular clamp is applied to the proximal common femoral artery. The vascular graft is removed in its entirety and the lumen of the common femoral artery inspected. Thrombotic or fibrinous debris is removed and intimal defects repaired. The common femoral arteriotomy is usually closed with a saphenous vein patch. Open removal is also recommended when there has been proximal percutaneous insertion in a morbidly obese patient. Open removal in this instance permits recognition of transperitoneal balloon placement and avoids the potential for intraabdominal bleeding following percutaneous balloon withdrawal. Open removal is also employed in patients who develop limb ischemia following percutaneous insertion. When the balloon is removed proximal and distal thrombectomy/embolectomy is performed with femoral artery repair as needed.

If the IABP has been inserted into the ascending aorta, the patient must be returned to the operating room for a repeat sternotomy. The IABP is carefully

Table 3.5. Indications for open intraaortic balloon removal

1. Open intraaortic balloon insertion.
2. Proximal percutaneous insertion in a morbidly obese patient.
3. Intraaortic balloon counterpulsation complicated by limb ischemia.
4. Transthoracic balloon insertion.

withdrawn through the vascular graft in an effort to avoid distal embolization of any fibrinous or thrombotic material that may have collected on the balloon or catheter. Depending upon the patient's condition and accessibility of the insertion site, the vascular graft can be removed from the ascending aorta and the aortotomy repaired directly. Alternatively, the vascular graft can be simply transected and oversewn leaving a remnant attached to the ascending aorta. If the IABP was inserted without a vascular graft the balloon is simply withdrawn and the pursestring suture tied. During the repeat sternotomy the mediastinum is copiously irrigated with an antibiotic solution. Routine perioperative antibiotic coverage should be provided.

COMPLICATIONS OF INTRAAORTIC BALLOON COUNTERPULSATION (TABLE 3.6)

The first problem associated with IABP counterpulsation is an inability to insert the balloon.[9] When cannulating the common femoral artery it is best to select the leg with the strongest pedal pulses. A J-tipped guidewire should be employed and if resistance is met as the balloon is advanced force should never be applied. An aortic dissection or aorto-iliac perforation characteristically occur at the time of IABP insertion and are documented in up to 1.5% of patients. An aortic dissection usually results in exsanguinating retroperitoneal hemorrhage although patients have been effectively counterpulsed with the IABP located within the aortic wall. Should an aortic dissection occur the balloon should be removed and, if necessary, reinserted via the opposite femoral artery utilizing guidewire and fluoroscopic guidance. Once the balloon is positioned in the descending thoracic aorta, its intraluminal location can be confirmed by withdrawing blood from the central lumen.

Vascular complications occur in up to 20% of patients.[10] Risk factors for developing a vascular complication following femoral IABP insertion include systemic hypertension, diabetes mellitus, female gender and peripheral vascular disease. The majority of vascular complications are thromboembolic in nature and occur most frequently when an IABP is inserted through the femoral artery. Arterial obstruction is usually the result of thromboemboli that originate on the balloon surface, cholesterol emboli that are dislodged from the vessel intima or mechanical obstruction due to a malpositioned IABP. If a patient's pedal pulses are lost during IABP counterpulsation, the balloon should be removed. Balloon removal usually results in a prompt return of distal pulses. If the limb remains ischemic

3

Table 3.6. Complications of use of an intraaortic balloon pump

Vascular
   Vessel wall injury
      Aortoiliac perforation
      Aortic dissection
   Ischemia
      Loss of distal pulse
      Thromboembolism
         Extremity
         Visceral
   Delayed complications
      False aneurysm
      Femoral artery stenosis
Neurologic
   Paresthesias
   Ischemic neuropathy
   Paralysis
Septic
   Fever
   Bacteremia
   Local wound infection
Hemorrhagic
   At insertion site
   Secondary to anticoagulation
Lymphocele
Balloon leak
   Air embolism
   Entrapped balloon
Improper balloon position

Reproduced with permission from: Richenbacher WE, Pierce WS. Management of complications of intraaortic balloon counterpulsation. In: Waldhausen JA, Orringer MB, eds. Complications in Cardiothoracic Surgery. St. Louis: Mosby-Year Book, Inc., 1991:97-102.

the patient should undergo femoral arterial exploration, thrombectomy and vein patch angioplasty. Rarely, a balloon dependent patient with an ischemic leg will require a femorofemoral crossover graft, the distal end of which is inserted into the proximal superficial femoral artery below the IABP insertion site (Fig. 3.12). Women, patients with a small body habitus or patients with known or suspected peripheral vascular disease are best served by a low profile sheathed balloon or sheathless balloon insertion. The smaller diameter catheter reduces the potential for vascular obstruction in a patient with diminutive femoral or iliac arteries. Systemic anticoagulation with heparin sodium affords the patient who does not require surgical intervention with some degree of protection from systemic embolization. Patients who require an open heart procedure or who have recently been separated from CPB do not require systemic anticoagulation during the period of IABP counterpulsation. CPB-induced platelet dysfunction protects the balloon surface and reduces the potential for systemic embolization.

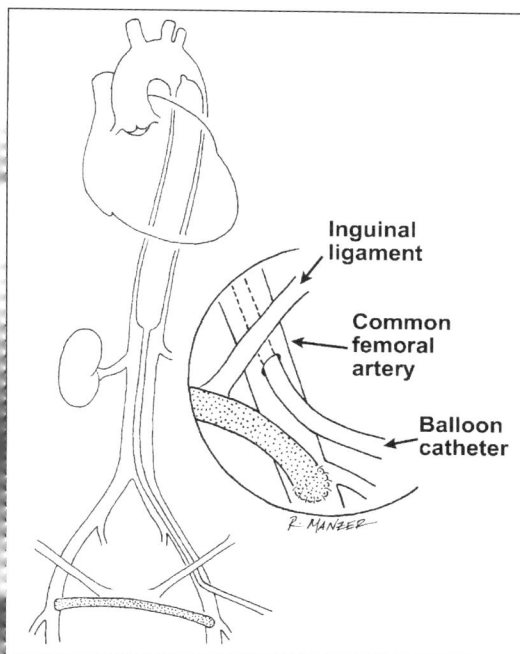

Fig. 3.12. Balloon dependent patients with leg ischemia may be managed with a femorofemoral crossover graft.

Inguinal ligament

Common femoral artery

Balloon catheter

Minor complications of IABP counterpulsation include bleeding at the insertion site, superficial wound infections, lymphocele or seroma formation. These complications occur infrequently and are usually self-limited or resolve following IABP removal. Minor bleeding at the insertion site usually ceases following discontinuation of heparin sodium therapy and correction of the patient's underlying coagulopathy. If it is felt that the patient would benefit from a low level of anticoagulation, balloon dependent patients who have not had or will not require CPB may be treated with low-molecular-weight dextran (10% dextran, 20 ml/hr IV).

Balloon rupture, a unique complication of IABP counterpulsation, is attributed to damage to the balloon at the time of insertion or contact between the balloon and a calcified atherosclerotic plaque.[11] The balloon shuttle gas, helium, can cause a fatal gas embolism in the event of a balloon leak. Balloon perforation is recognized by failure to achieve diastolic augmentation and the presence of blood in the balloon shaft. Counterpulsation should be immediately discontinued. The patient is placed in steep Trendelenburg position and suction applied to the balloon catheter. The balloon is removed and replaced. Neurologic deficits attributable to gas emboli are uncommon but potentially treatable with hyperbaric oxygen therapy.

Transthoracic balloon insertion has a complication rate that is significantly less than that which occurs when the IABP is inserted via the femoral artery. The difference in complication rates between the two insertion sites is attributable to the virtual elimination of vascular problems when the IABP is inserted into the ascending aorta when compared to those encountered with femoral arterial cannulation. Complications peculiar to cannulation of the ascending aorta include graft infection with mediastinitis, coronary artery or cerebrovascular embolization and mechanical tamponade of mediastinal structures. Should mediastinitis occur despite strict sterile technique and perioperative antibiotic coverage the balloon and adjacent prosthetic material must be removed. If a side-arm graft has been employed on the ascending aorta the graft should be removed in its entirety and the ascending aorta debrided and repaired. Brief occlusion of the carotid arteries and coronary artery bypass grafts, if present, reduces the potential for embolization to these vessels during IABP removal. Mechanical compression of mediastinal structures may preclude sternal closure at the time of balloon insertion. In particular, it is important to orient the IABP away from vein grafts due to the potential for graft compression and myocardial ischemia. If the balloon causes cardiac or vein graft compression when the sternum is approximated, the sternum should be stented open and the skin closed. Alternatively, the IABP may be brought through the chest wall via the right second intercostal space.

REFERENCES

1. Bregman D, Kaskel P. Advances in percutaneous intra-aortic balloon pumping. Crit Care Clin 1986; 2:221-236.
2. Arafa OE, Pedersen TH, Svennevig JL et al. Intraaortic balloon pump in open heart operations: 10-year follow-up with risk analysis. Ann Thorac Surg 1998; 65:741-747.
3. Creswell LL, Rosenbloom M, Cox JL et al. Intraaortic balloon counterpulsation: Patterns of usage and outcome in cardiac surgery patients. Ann Thorac Surg 1992; 54:11-20.
4. Naunheim KS, Swartz MT, Pennington DG et al. Intraaortic balloon pumping in patients requiring cardiac operations. Risk analysis and long-term follow-up. J Thorac Cardiovasc Surg 1992; 104:1654-1661.
5. Dietl CA, Berkheimer MD, Woods EL et al. Efficacy and cost-effectiveness of preoperative IABP in patients with ejection fraction of 0.25 or less. Ann Thorac Surg 1996; 62:401-409.
6. Aguirre FV, Kern MJ, Bach R et al. Intraaortic balloon pump support during high-risk coronary angioplasty. Cardiology 1994; 84:175-186.
7. Santini F, Mazzucco A. Transthoracic intraaortic counterpulsation: A simple method for balloon catheter positioning. Ann Thorac Surg 1997; 64:859-860.
8. Mertlich, G, Quaal SJ. Air transport of the patient requiring intra-aortic balloon pumping. Crit Care Nurs Clin of North Am 1989; 1:443-457.
9. Richenbacher WE, Pierce WS. Management of complications of intraaortic balloon counterpulsation. In: Waldhausen JA, Orringer MB eds. Complications in Cardiothoracic Surgery. St. Louis: Mosby-Year Book, Inc., 1991:97-102.
10. Gol MK, Bayazit M, Emir M et al. Vascular complications related to percutaneous insertion of intraaortic balloon pumps. Ann Thorac Surg 1994; 58:1476-1480.
11. Sutter FP, Joyce DH, Bailey BM et al. Events associated with rupture of intraaortic balloon counterpulsation devices. ASAIO Trans 1991; 37:38-40.

Extracorporeal Membrane Oxygenation (ECMO) Support for Cardiorespiratory Failure

Ralph E. Delius, Angela M. Otto 4

INTRODUCTION

Extracorporeal membrane oxygenation (ECMO) has been found to be a life-saving therapy in selected groups of patients. Although initially met with skepticism since it was first performed by Hill and subsequently popularized by Bartlett,[1] ECMO has been proven to be useful and effective therapy in the management of neonatal respiratory failure.[2,3] Encouraged by the success of ECMO in this population, ECMO has been increasingly applied to other groups of patients with cardiorespiratory failure including pediatric and adult patients with respiratory failure, children with cardiac failure and even as a means of resuscitation.[1,4,5]

Over 12,000 neonates worldwide have received ECMO support for respiratory failure. The overall survival rate has been 83% with survival exceeding 90% in certain diagnostic subgroups. These survival rates are impressive considering that these patients had an estimated mortality risk of 80% prior to initiation of ECMO. The role of ECMO support in pediatric and adult patients with respiratory failure has not been established unequivocally but retrospective reports suggest that ECMO may be useful in these groups of patients as well.[5,6] Approximately 50% of pediatric and adult patients placed on ECMO for respiratory failure have survived to discharge.

ECMO for cardiac support has largely been limited to pediatric patients.[4,7] Adults with cardiac failure have other options including ventricular assist devices (VADs) and the intraaortic balloon pump (IABP). These approaches have a somewhat limited application to the pediatric population in part because of the lack of size-appropriate equipment. ECMO support can be used in almost any size of patient and has been used extensively for cardiac support in pediatric patients. The largest application is in patients with cardiac failure following congenital heart surgery. Approximately 40-50% of these patients survive to discharge.[4,7] ECMO has also been used in a more limited application as a bridge to transplantation for pediatric patients with end stage heart disease.

TEAM CONCEPTS

ECMO is best performed by an organized and established team. Impromptu ECMO, often run with local or proprietary perfusionists with only a limited background in ECMO support may occasionally be successful but more often than not is a recipe for disappointment. From an economic standpoint, approximately 10-12 cases per year will be necessary to justify the expenditures for equipment, salary for an ECMO coordinator and overtime. Very few ECMO centers can support this volume by limiting their focus to postcardiotomy patients. Consequently, almost all successful ECMO programs provide support for neonatal respiratory failure (by far the most widespread application of ECMO) in addition to other patient groups that require ECMO support relatively infrequently.

ECMO specialists are responsible for the minute to minute management of the ECMO circuit. They are typically drawn from the ranks of respiratory therapists, intensive care unit (ICU) nurses and occasionally perfusionists. Their duties include controlling pump flow, regulating anticoagulation and monitoring pump pressures. In most programs there is one ECMO specialist per patient but in some busy programs an ECMO specialist may be responsible for two patients. The ECMO specialist assists the ICU nurse primarily responsible for the patient and operates under the direction and supervision of a licensed physician.

The ECMO team is usually directed by an ECMO coordinator, whose responsibilities are legion even in smaller programs. As a consequence, their efforts are usually directed full time to the ECMO program. ECMO coordinators typically have extensive experience in critical care, respiratory therapy or CPB perfusion. Typical responsibilities of the ECMO coordinator include initial training, organizing refresher courses two to three times per year, maintaining equipment and supplies, serving as a troubleshooter and also filling in as an ECMO specialist when gaps in the call schedule occur. Simply maintaining a call schedule for ECMO can absorb a large amount of time particularly when the ECMO specialists are drawn from several different administrative units within the hospital (respiratory therapy, perfusion, neonatal and pediatric ICUs).

Support from the perfusionists who are primarily involved with CPB for operative support can be very helpful particularly if equipment is shared. Perfusionists can provide technical support and can be helpful with troubleshooting equipment problems as well.

Careful coordination of ancillary support is imperative and should be organized at the time an ECMO team is created. Surgeons, neonatologists and pediatric intensivists will be involved in any ECMO endeavor. Arrangements will need to be made with the blood bank as ECMO can be an extensive drain on blood bank resources. Furthermore, the indications for transfusion of patients on ECMO support are considerably different from other groups of patients. Clarification of needs and indication beforehand can prevent squabbles over needed blood products when a critical situation arises.

Some ECMO patients also require renal support such as hemofiltration or dialysis. Involvement of a nephrologist with an interest in acute renal failure can be very helpful. Contact should be made early in any ECMO run in which dialysis may be needed.

Full support from hospital administration is needed before embarking on starting an ECMO program. Initial cash outlay is needed for equipment, training and hiring a coordinator. Overtime pay may be necessary for the ECMO specialists. Arranging the call schedule for the ECMO specialists is easier when there is full support of the administration and can prevent disagreements between the various units from which the ECMO specialists are drawn.

Physician direction of the ECMO program is integral to the success of any program. Directors typically have prior expertise in the use of ECMO and can be drawn from surgeons, intensivists or neonatologists. In some hospitals ECMO is primarily run by neonatologists with pediatric or cardiothoracic surgeons providing only surgical support, while in other programs surgeons are primarily responsible for the ECMO program with neonatologists and intensivists playing key roles by assisting with patient management. Both models have worked effectively.

ECONOMIC CONSIDERATIONS

High healthcare costs have been a major concern for the last 8-10 years. Particularly implicated in the escalation of healthcare costs has been high technology interventions typically provided in an intensive care setting. ECMO is a classic example of a high cost intervention. Concerns about cost effectiveness and efficacy delayed widespread acceptance of ECMO in the United Kingdom until prospective randomized trials conclusively proved its usefulness.[2] Financial considerations will continue to be a primary consideration for any new or established ECMO program.

ECMO PROGRAM DEVELOPMENT COSTS

The first step in determining whether an ECMO program is needed is examining regional need. Approximately 12 patients per year will be needed to maintain clinical skills; this also approximates the number of patients required for financial viability of a program, although this is greatly dependent on payor mix. It has been estimated that 1 neonate out of every 3,000 live births requires ECMO support for neonatal respiratory failure. Approximately 1-3% of pediatric patients undergoing open heart surgery require ECMO support postoperatively. A considerably smaller number will require ECMO support preoperatively for myocarditis or as a bridge to transplantation. The number of pediatric and adult patients requiring ECMO support is impossible to predict at this point in time.

The estimated cost of starting a new ECMO program is $125,000-$150,000. Most expenses are related to equipment and training. Cost centers for a developing program are shown in Table 4.1.

OPERATING COSTS

Operating costs are mostly related to personnel salaries although equipment replacement should also be considered. Other costs include updates, training drills and periodically training new ECMO specialists as the need arises. Travel costs to ECMO meetings should also be factored into operating costs.

CHARGES

International Classification of Diseases, Ninth Revision, Clinical Modification (ICD-9-CM) codes are listed in Table 4.2. Procedures performed on the ECMO patient are routinely coded although adding a modifier for increased difficulty

---

Table 4.1. Extracorporeal membrane oxygenation (ECMO) program cost centers-program development

---

ECMO equipment (minimum of two circuits)

ECMO coordinator salary

ECMO specialist wages during training

Office space

Telephone

Printing costs (training manuals)

Travel to established ECMO center for consultation

Laboratory training

    Animals

    Blood product

    Laboratory space

    Disposables—ECMO circuit, ACT tubes, etc.

ACT = activated clotting time

Table 4.2. *International classification of diseases, ninth revision, clinical modification (ICD-9-CM) codes*

	Procedure	ICD-9-CM code
ECMO	cannulation	36822
ECMO	decannulation	33978
ECMO	decannulation with vessel repair	35201
First day care		33960
Subsequent daily care		33961

ECMO = extracorporeal membrane oxygenation

due to the anticoagulation needed and technical difficulties related to ECMO support can be justified.

CIRCUIT DESIGN

ECMO circuits are similar but not identical to circuits used for open heart surgery. In contrast to perfusion circuits for open heart surgery which are open (have a reservoir of blood or fluid), ECMO circuits are closed (no reservoir is present). The amount of blood flowing out of the patient is identical to the volume of blood simultaneously reinfusing back into the patient.

A standard ECMO circuit is shown in Figure 4.1. The venous blood is drained through a cannula that is usually placed in the right atrium via the internal jugular or common femoral vein. Blood then flows into a distensible bladder which serves as an important safety device. If pump flow exceeds venous drainage the bladder will collapse. The volume or pressure within the bladder is continuously monitored. If collapse of this bladder is detected pump flow is automatically stopped or slowed. The bladder then refills and normal pump operation resumes.

From the bladder the blood passes through a blood pump which provides kinetic energy to move blood through the circuit and into the patient. Most programs prefer a roller pump which is simple and causes minimal hemolysis. Some programs have successfully used centrifugal pumps in older patients. This pump is perhaps safer than the roller pump in that it is demand regulated and less likely to pump air. However, in ECMO circuits these pumps appear more likely to cause hemolysis and therefore have a somewhat limited application.

The blood then flows from the pump to a membrane oxygenator. Currently only one oxygenator is Food and Drug Administration (FDA) approved for ECMO use (Medtronic Perfusion Systems, Brooklyn Park, MN). Hollow fiber oxygenators

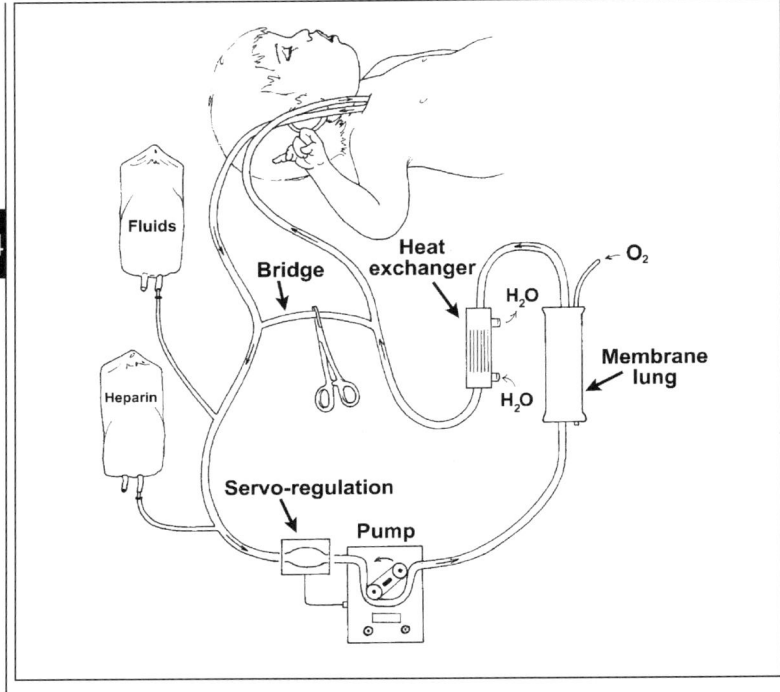

Fig. 4.1. Diagram of venoarterial ECMO circuit.

which are widely used in operative circuits have not been proven to be useful for long-term support. The size of the membrane lung is critical. The membrane lung should be capable of providing complete cardiopulmonary support. Each size of the membrane lung has a rated flow which specifies the maximum flow of normally saturated venous blood which will leave the oxygenator 95% saturated. The membrane lung chosen for any given patient should have a rated flow equivalent to or in excess of the cardiac output of the patient. An excessively large oxygenator is not desirable either as priming volume will be increased and blood flow through the pump will be sluggish promoting the likelihood of thrombosis. Recommended tubing and membrane lung sizes are shown in Table 4.3.

Blood then flows from the membrane lung to the heat exchanger which serves to heat the blood before returning it to the patient. Several satisfactory products are on the market all of which work on the countercurrent flow principle. A pump heats and circulates water through the heat exchanger.

The oxygenated blood is then returned to the patient via an arterial (for venoarterial support) or venous (for venovenous support) cannula. A bridge that

Table 4.3. Recommended raceway, tubing and oxygenator sizes

Weight (kg)	Raceway	Tubing	Oxygenator (m²)
2-10	1/4"	1/4"	0.8
11-19	3/8"	1/4"	1.5
20-34	1/2"	3/8"	2.5
35-69	1/2"	1/2"	3.5
70-95	1/2"	1/2"	4.5*

* Two oxygenators connected in parallel may be needed for venovenous bypass

connects the venous outflow and arterial inflow allows recirculation of blood while the patient is off bypass.

LOCATION

ECMO support is usually initiated in the ICU except for those patients placed on ECMO in the operating room following open heart surgery. Battery support for the ECMO circuit is essential for transfer of this group of patients back to the ICU. A battery pack is also essential when a patient on ECMO in the ICU requires transfer to the radiology department, cardiac catheterization laboratory or elsewhere in the hospital.

A place in the ICU is chosen that will accommodate the pump, ECMO specialists, operating room team during cannulation and equipment cart. If possible either the largest cubicle or if in an open intensive care unit setting two spaces should be set aside for an ECMO patient.

Once patients are cannulated they are difficult to move so they should be placed in an appropriate bed and carefully positioned before cannulation is initiated. Easy access to cannulation sites and the patient is essential for nursing and surgical staff. In our unit, this usually means placing the head of the patient away from the wall and monitors opposite the usual position. The ECMO pump is initially placed to the right or left of the patient during cannulation but then is moved to the head of the bed after cannulation has been completed. This is very easy to accomplish and allows complete access to the patient. Subsequent procedures are preferentially performed in the ICU rather than the operating room. There are occasions where this is not feasible, for example in a patient requiring correction of congenital heart disease. Nevertheless, transfers around the hospital should be minimized if possible since moving the patient is time consuming, requires multiple personnel and carries a potential risk.

VENOVENOUS VERSUS VENOARTERIAL BYPASS

The most widely used form of ECMO is venoarterial (VA) ECMO which involves cannulation of both the right common carotid artery and internal jugular vein. One of the initial objections to ECMO was the accompanying ligation of the common carotid artery although there has been no conclusive evidence that this practice has been harmful. Nevertheless, persistent concerns about common carotid artery ligation led to the development of a double lumen cannula suitable for neonates which allows for drainage of the venous blood and reinfusion of oxygenated blood.[8] Older patients have several other options for venous access and typically drainage occurs through a cannula placed in the right jugular vein and reinfused through a smaller cannula placed in the femoral vein. Venovenous (VV) ECMO is reserved for patients that are hemodynamically stable. The definition of hemodynamic stability varies from observer to observer but in general if the primary pathophysiologic problem is respiratory failure and only modest doses of inotropes are required (dopamine hydrochloride < 10 mcg/kg/min, dobutamine hydrochloride < 20 mcg/kg/min, epinephrine < 0.1 mcg/kg/min) then VV ECMO should be seriously considered. In addition to avoiding arterial ligation, VV ECMO also has several other theoretical advantages (Table 4.4).

PHYSIOLOGIC CONSIDERATIONS

Reinfusion of oxygenated blood into the venous side of the circulation is somewhat physiologically analogous to congenital heart patients with a single ventricle. The admixture of desaturated blood on the venous side of the circulation leads to a lower arterial oxygen saturation (SaO_2) compared to when VA perfusion is used. If no native lung function is present the SaO_2 in the main pulmonary artery will be identical to the systemic SaO_2. If the systemic SaO_2 is greater than the SaO_2 in

---

Table 4.4. Venovenous bypass

Advantages
    Avoids arterial cannulation
    Normal pulse pressure
    Lungs can serve as microemboli filter *
    Normal blood flow through lungs may enhance reparative process *
Disadvantages
    No hemodynamic support

---

* Theoretical advantage

the main pulmonary artery then some contribution by the native lung must be present. As in patients with cyanotic congenital heart disease, moderate degrees of systemic desaturation (SaO_2 80-90%) are well tolerated provided cardiac output and oxygen carrying capacity are optimized.

Another physiologic feature unique to VV bypass is the concept of recirculation. Some of the reinfused oxygenated blood is immediately drained out of the right atrium and back in the ECMO circuit without ever being delivered to the systemic side of the circulation. Recirculation creates an inefficiency in the system since the oxygen content of the venous blood draining into the ECMO circuit is elevated. Less oxygen can therefore be added to any given volume of blood. This can be overcome simply by increasing flow across the oxygenator. This factor accounts for why patients on the VV support require 25-50% greater pump flow rates than patients supported with VA ECMO.

Occasionally patients will fail VV bypass due to subsequent development of hemodynamic instability or rarely inadequate oxygenation. An arterial cannula is then placed in the common carotid artery. The previously inserted reinfusion cannula is then converted to a drainage cannula by attaching this cannula to the drainage arm of the circuit by means of a Y-connector. Once the arterial cannula has been inserted, conversion from VV to VA ECMO can be accomplished easily in 15-20 seconds.

INDICATIONS FOR ECMO SUPPORT

NEONATES

Most neonates considered for ECMO support have respiratory failure with pulmonary hypertension of the newborn often being the underlying mechanism of respiratory distress, irrespective of the primary insult. Medical management of this group of neonatal patients has recently undergone revolutionary changes with the introduction of inhaled nitric oxide and high frequency oscillatory ventilation. Despite these new advances occasionally patients will fail this therapy and require ECMO support. The oxygenation index (OI) which was derived during the so-called conventional ventilation era is still applicable despite these new advances. The OI is defined as:

$$OI = \frac{\text{mean airway pressure x } FiO_2 \text{ x } 100}{PaO_2}$$

where FiO_2 is the inspired oxygen fraction and PaO_2 is the arterial oxygen tension. An OI > 40 defines an 80% mortality risk even if nitric oxide and high frequency oscillatory ventilation are used. A neonate with an OI > 40 based on three of five post ductal arterial blood gases drawn approximately 30 minutes apart are considered candidates for ECMO. Some institutions prefer to use the alveolar-arterial oxygen difference (AaDO$_2$) as an indication for ECMO although the oxygenation index is probably more widely used. The formula for AaDO$_2$ gradient is:

$$AaDO_2 = 760 - (P_aO_2 + P_aCO_2 + 47)$$

where PaCO$_2$ is the arterial carbon dioxide tension. A value consistently over 610 has been equated to an 80% mortality risk.

It is necessary to define and exclude from ECMO support patients with no reasonable chance of recovery. Patients with an estimated gestational age less than 35 weeks are usually excluded as the risk of intercerebral hemorrhage is increased in this younger group of patients. If possible, all neonates should be screened for congenital heart disease and intracranial hemorrhage prior to initiating ECMO support. Patients with congenital heart disease should be managed based on the physiology and anatomy of the heart defect. Patients with grade 1 intracranial hemorrhage have been successfully supported with ECMO with modified antico-agulation regimens. However, more severe grades of intracranial hemorrhage are usually excluded.

RESPIRATORY FAILURE IN ADULTS AND OLDER CHILDREN

The benefit of ECMO support in this group of patients has not been definitely proven in clinical trials. Two trials have failed to show any benefit from ECMO in adult patients but both trials had significant concerns that may have affected the conclusions of the studies.[9,10] Until recently there has been little significant im-provement in the clinical outcome of patients with life threatening respiratory failure. However, recent reports have suggested an overall improvement in sur-vival in this challenging group of patients and it appears that incremental changes in care appear to be steadily improving results. As such, the role of ECMO re-mains uncertain.

The present indications for ECMO support are a transpulmonary shunt greater than 30% despite optimal ventilation and pharmacological management and a static compliance less than 0.5 ml/cmH$_2$O/kg. Separating patients with reversible lung disease from those with irreversible disease remains a challenge. In adults and older children the diseases that commonly cause respiratory failure lead to interstitial inflammation which in turn often leads to fibrosis or necrosis. Most groups restrict ECMO support to patients in the acute phase of their illness with a maximum of 5-7 days of mechanical ventilation prior to ECMO support. In the younger patient contraindications to ECMO support include irreversible dis-ease, neurological injury, terminal malignancy or any contraindication to anticoagulation.

ECMO FOR CARDIAC SUPPORT

Most patients in this category are patients with cardiac failure following surgi-cal repair although other clinical scenarios such as myocarditis or support until a donor heart becomes available are not unusual. In the older patient other options such as IABP or VAD are available and are usually preferable to ECMO. In younger patients, however, options are limited. Furthermore, cardiac failure is often biventricular in this group of patients and there is often a component of respiratory failure as well.

Indications for ECMO in this group of patients have not been well delineated. The most obvious indication is failure to wean from bypass following open heart surgery. However, it is absolutely mandatory to rule out an anatomic or technical problem prior to considering ECMO support. This consideration is also true in patients who are considered for ECMO support several hours following surgery. Simpler measures, such as opening the chest of postcardiotomy patients should be considered before embarking on an ECMO course. However, if patients continue to deteriorate despite opening of the chest and maximal medical therapy, ECMO should be considered if the cardiac output index is persistently less than 2 L/min/m^2, mixed venous oxygenation is consistently less than 50% (for biventricular failure) or if a persistent metabolic acidosis is present for greater than 2-4 hours.

CANNULATION TECHNIQUES

NEONATES

Virtually all neonates with respiratory failure are cannulated through the right internal jugular vein and common carotid artery. Positioning of the patient is particularly important. A roll is placed behind the shoulders and the head rotated to the left (Fig. 4.2). Morphine sulfate (0.1 mg/kg) is used for analgesia and 1-2 ml

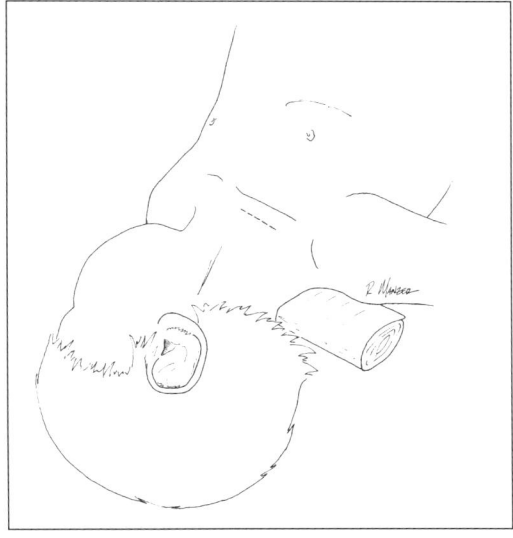

Fig. 4.2. Proper positioning of the patient prior to cervical cannulation.

of 1% lidocaine hydrochloride is given as a local anesthetic. Pancuronium bromide (0.1 mg/kg) is given to prevent spontaneous breathing during cannulation which can lead to air embolism during venous cannulation. Either a transverse incision or one along the anterior border of the sternocleidomastoid muscle is used. The dissection is carried down between the two heads of the sternocleidomastoid muscle (Fig. 4.3). The internal jugular vein is usually easily located. The right common carotid artery is located medial and slightly posterior. The omohyoid muscle may need to be divided to completely expose the artery and vein. The vagus nerve resides posteriorly between these two vessels and should be avoided (Fig. 4.4). Heparin sodium (100 U/kg) is given after dissection has been completed. The artery is dissected out even if VV ECMO is contemplated in case emergent conversion to VA ECMO becomes necessary (Fig. 4.5). A 2-0 silk tie is placed on the arterial and venous cannulae at approximately 2.5 and 5.5 cm respectively to serve as a marker for the desired length of cannula to be inserted. The two ends of the ties should be left long as they will serve to anchor the cannulae later. The artery and vein are then ligated distally with a 2-0 silk ligature. The artery is usually cannulated first. Appropriately sized vascular clamps are placed and a small incision is made in the vessel. To prevent dissection while inserting the cannula one or two 7-0 polypropylene sutures are placed across the proximal wall of the artery. This prevents the intima from sliding into the artery while the cannula is inserted (Fig. 4.6). Once the cannula is inserted the previously placed tie around the vessel is secured over a short segment of vessel loop. The vessel loop serves to protect the underlying vessel wall when the ties are later removed (Fig. 4.7). This

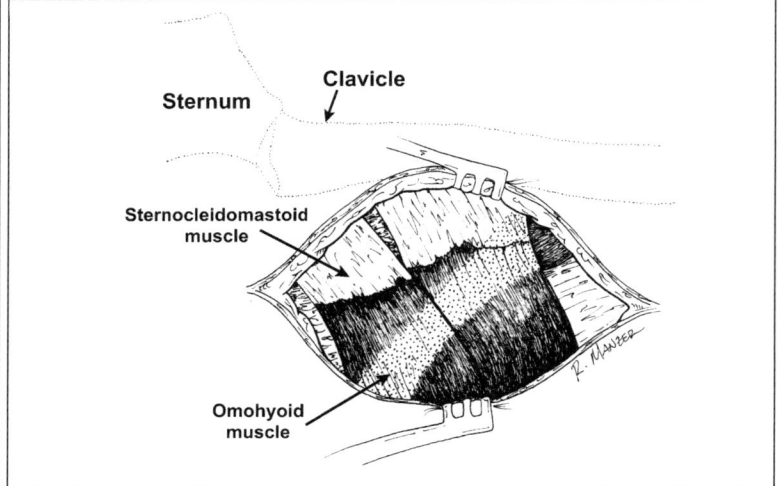

Fig. 4.3. Two heads of the sternocleidomastoid muscle and the omohyoid can be clearly seen through a transverse surgical incision.

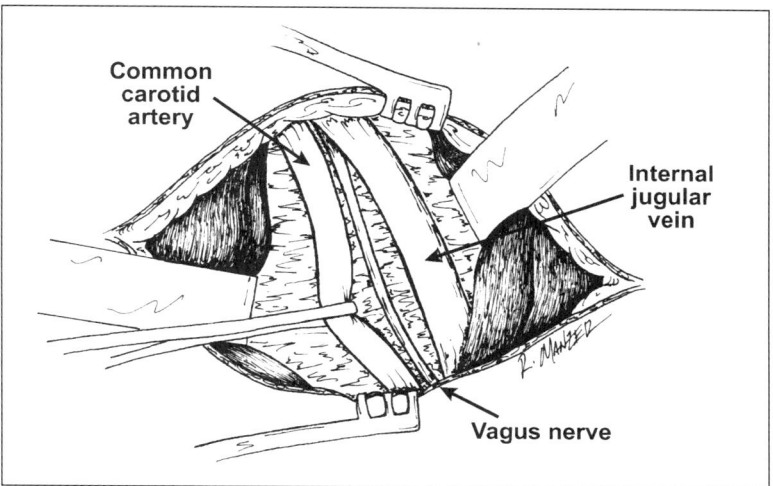

Fig. 4.4. Exposure of the internal jugular vein and common carotid artery.

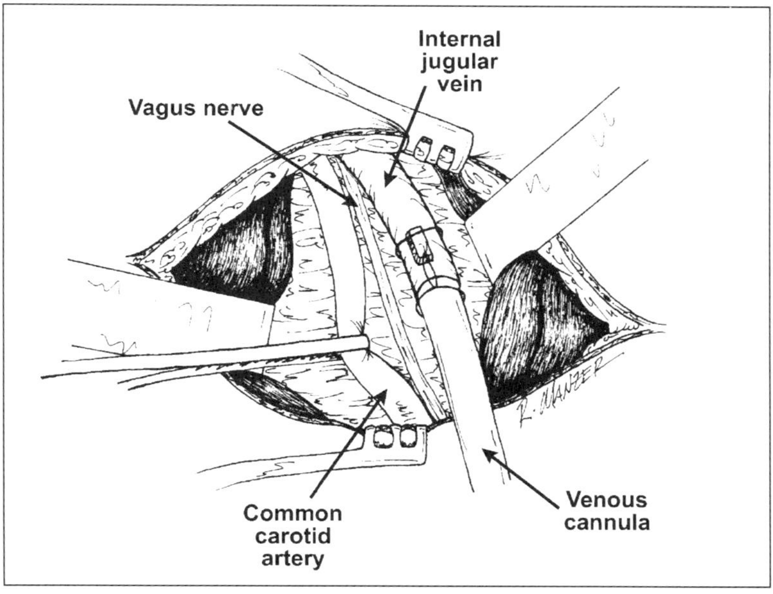

Fig. 4.5. Cannulation of the right internal jugular vein. Note the segment of vessel loop underneath the ties.

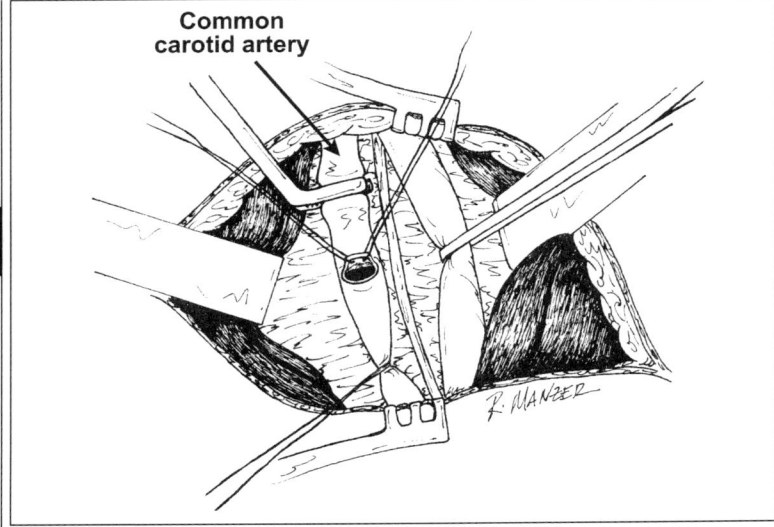

Fig. 4.6. Stay sutures fixed the intima to prevent dissection.

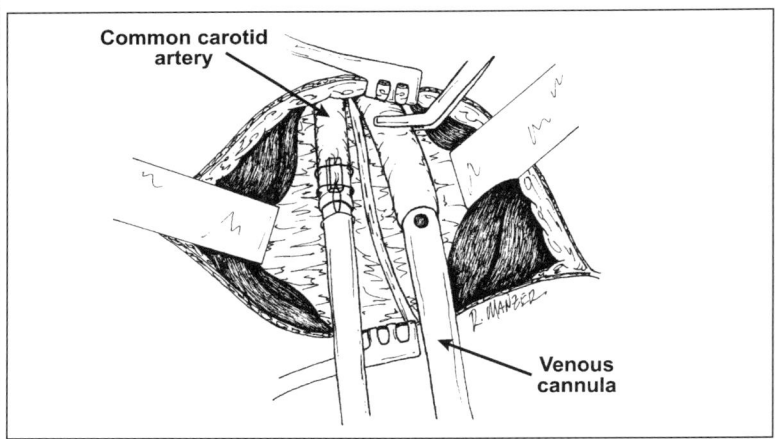

Fig. 4.7. Cannulation of the right internal jugular vein. Note that the artery is cannulated first for venoarterial bypass.

same tie is then secured to a tie that was previously placed around the cannula and helps secure the cannula in place. The cannula is then tightly secured to the skin (Fig. 4.8). The venous cannulation is performed in a similar manner although 7-0 polypropylene tacking suture is usually not necessary. If VV bypass is chosen, correct orientation of the cannula must be assured by placing the inflow (arterial)

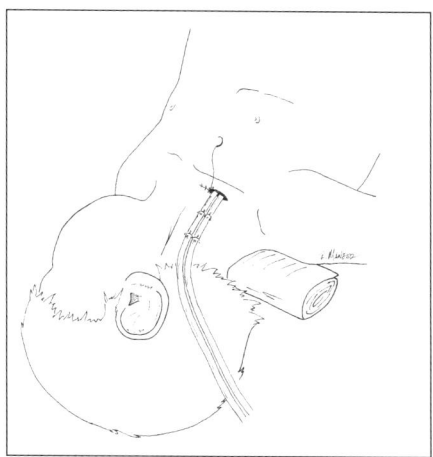

Fig. 4.8. Closure of the cannulation incision.

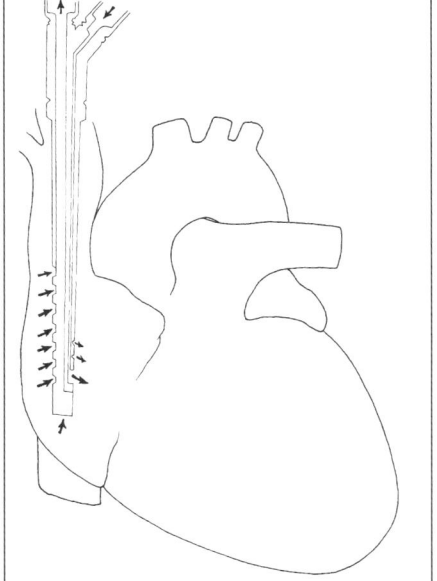

Fig. 4.9. Diagram of double lumen cannula. Note that the reinfusion ports are directed anteriorly towards the tricuspid valve.

4

ports anteriorly (Fig. 4.9). This will minimize recirculation with VV bypass and direct the arterial return through the tricuspid valve. Hemostasis is obtained after the cannulae are secured and attached to the ECMO circuit. Some groups prefer to place fibrin glue in the wound at this point in time. We prefer to save the fibrin glue as an easy remedy if bleeding later develops during the ECMO course. Once hemostasis has been obtained the wound is then closed with a running 4-0 nylon suture. A chest roentgenogram is then immediately obtained to assess cannulae placement (Fig. 4.10).

OLDER CHILDREN AND ADULTS

Cannulation in older patients usually involves placing two cannulae since VV bypass is usually preferred in this group of patients. If VA bypass is necessary it is still preferable to cannulate the common carotid artery. Although this raises concerns about potential cerebral injury, in fact injury is rare unless atherosclerotic disease is present. Cannulating the femoral artery may seem preferable but the heart will continue to eject poorly oxygenated blood creating a situation where the upper half of the body is cyanotic while the lower half is well oxygenated. If VA bypass is necessary due to hemodynamic instability, it may be preferable to consider later conversion to VV bypass once the patient is hemodynamically stable.

VV cannulae may be placed percutaneously or by surgical cutdown. Usually the venous drainage cannula is placed in the right atrium via the femoral vein or the right internal jugular vein while a shorter return cannula is placed in the opposite femoral vein. Some groups have had excellent success with percutaneous placement of both cannulae. Percutaneously placed venous drainage cannulae seem to have less bleeding than surgically placed cannulae. However, the complication rate seems to be higher and as a consequence we prefer the surgical approach. In the hands of an experienced surgeon the differences between the two techniques in time required for cannulation are quite minimal. Cannulation of the internal jugular vein is identical to the technique in neonates. The return cannula, which is

Fig. 4.10. Correct placement of the arterial and venous cannulae.

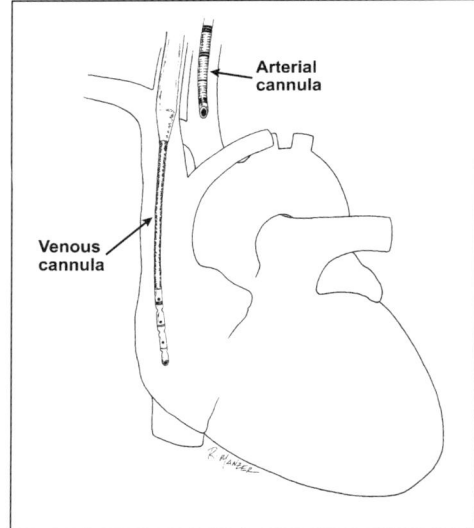

often much smaller, can be safely placed percutaneously. Our preference is to place the return cannula in the femoral vein using the Seldinger technique. The Bio-Medicus cannulae (Medtronic Bio-Medicus, Inc., Eden Prairie, MN), which are wire wound and very thin walled, have excellent flow characteristics and are our preference. Usually a 17 or 19 F reinfusion cannula will suffice for most adult patients. Regardless of the approach chosen, a 21-25 F venous drainage cannula is usually required in this group of patients.

CARDIAC SUPPORT

4

ECMO for cardiac support is by definition always VA bypass. Patients placed on ECMO support in the operating room will typically use the operative cannulae with conversion consisting simply of disconnecting the operative bypass tubing and connecting it quickly to the ECMO circuit. Only one venous cannula is necessary.

For patients who come out of the operating room and subsequently develop cardiac failure several hours later the choices are either the cervical approach as used in patients with respiratory failure or transthoracic cannulation. The advantages and disadvantages of each are shown in Table 4.5. Cervical cannulation, as noted earlier, is identical to that described for neonates. Transthoracic cannulation is identical to that performed for operative bypass. It is critical in this group of patients to determine if left ventricular ejection is present particularly in light of the increased afterload caused by ECMO. If the left ventricle is not ejecting and the ventricular septum is intact pulmonary venous pressures may increase and cause pulmonary edema. Decompression of the left side of the heart must be performed by either balloon or blade arterial septostomy or placement of an operative left ventricular vent usually via the left atrium or right superior pulmonary vein. The left ventricular vent is then connected to the venous drainage line allowing gravity drainage and decompression of the left ventricle.

Table 4.5. Transthoracic versus cervical cannulation

	Advantages	Disadvantages
Transthoracic	Easy to vent left ventricle	High incidence of bleeding.
	More appropriate for complex venous anatomy	Increased risk of mediastinitis.
	Avoids carotid artery ligation	
Cervical	Simple	Requires carotid artery ligation.
	Minimal bleeding	Not applicable in complex venous anatomy.
	Low risk of infection	Difficult to vent left ventricle.

PATIENT MANAGEMENT

ANTICOAGULATION

Activated clotting times (ACT) are used to monitor heparin sodium administration during bypass. The ACT is monitored hourly at the bedside and the heparin sodium infusion rate (usually 30-60 U/kg/hr) is adjusted accordingly. In most cases, the ACT is maintained between 180-200 seconds. This provides satisfactory anticoagulation without needlessly increasing the risk of bleeding. There are circumstances in which the anticoagulation may be run at lower rates, such as in postoperative patients, although it must be realized that this increases the risk of clotting within the ECMO circuit. Usually the ACT can be safely maintained at 160-180 seconds providing full flow is maintained to minimize stasis within the circuit. In dire circumstances the ACT may be maintained at even lower levels although the risk of circuit clotting increases considerably. ECMO has even been performed without heparin administration, although the life span of the ECMO circuit is usually only 6-24 hours. If extremely low levels of anticoagulation are indicated ECMO pump flows must be maintained at the maximal achievable levels. It is also imperative to have a spare primed circuit available at the bedside which can be quickly recruited into service if the primary circuit clots.

PLATELETS

In a standard case, platelet counts are kept greater than 100 k/mm³ which usually is sufficient in patients with respiratory failure. If patients are at risk for bleeding complications, however, it has been empirically demonstrated that maintaining platelet counts > 150 k/mm³ is useful. Platelets are always transfused into the port distal to the oxygenator to prevent clumping within the oxygenator. Usually 10-15 ml/kg are administered over 15-30 minutes.

FIBRINOGEN

Fibrinogen levels are maintained > 100 mg% in most instances although in postoperative patients this may be increased to > 150 mg%. Cryoprecipitate (5-10 ml/kg) is usually administered into the port immediately prior to the bladder or can be given directly into the patient.

RED BLOOD CELLS

The hematocrit is fastidiously maintained at 40-45% to maximize oxygen delivery. In selected VV ECMO cases the hematocrit may even be preferably maintained at 45-50%. Anemia is poorly tolerated in these patients and should be avoided. Studies that suggest moderate anemia in critically ill patients is well tolerated are often quoted by blood bank directors; however, findings of these studies are not necessarily applicable to ECMO patients. Even a hematocrit of 38% should be treated. Packed red blood cells (10-15 ml/kg) are administered into the port proximal to the bladder.

Serum hemoglobin levels are monitored every 12 hours to detect hemolysis. If serum hemoglobin levels exceed 50 mg/dl, a search for causes of hemolysis should be initiated. The raceway occlusion should be checked for excessively tight occlusion. Extensive clot within the oxygenator, often manifest by increasing transoxygenator pressures, can also lead to hemolysis and may indicate that an oxygenator change is necessary.

FRESH FROZEN PLASMA
Fresh frozen plasma (FFP) can be administered for volume expansion if the hematocrit is in the desired range. In most cases, however, 5% albumin is preferable for volume expansion. FFP is usually reserved for patients with sepsis, disseminated intravascular coagulation, or replacement of deficient clotting factors. FFP (10-15 ml/kg) is given to the patient or into the port prior to the bladder.

BLEEDING
Strategies for bleeding involving lower levels of anticoagulation have been mentioned earlier. Useful adjuncts, particularly in postoperative patients, include aminocaproic acid (100 mg/kg bolus infused over 1 hour followed by 30 mg/kg/hr intravenous infusion). Antithrombin III (50-100 U/kg IV QD x 3 days) appears to minimize clotting within the circuit when lower levels of anticoagulation are used. Bleeding from cervical cannulation sites can be controlled by injecting fibrin glue (10 ml thrombin and 10 ml cryoprecipitate in separate syringes). Blunt needles are attached to these syringes and are inserted into the wound between the sutures. Approximately 5 ml from each syringe is injected in the wound. If bleeding persists, surgical exploration is necessary.

PUMP FLOW RATES
ECMO pump flows are adjusted according to whether the patient is on VA or VV bypass, patient needs and level of anticoagulation. As mentioned earlier, if lower levels of anticoagulation are chosen then flows are maintained at the maximum possible flow to minimize stasis. For most patients 80-100 ml/kg is satisfactory for patients supported with VA bypass while 100-130 ml/kg is necessary for VV support due to recirculation. Once the steady state has been initiated, flows are adjusted based on the mixed venous oxygen saturation (SvO_2). For VA supported patients the SvO_2 is kept > 65-70%. It is difficult to identify an ideal SvO_2 for patients on VV support since the degree of recirculation can vary from patient to patient. Usually a SvO_2 > 75-80% is desirable in VV patients. An SvO_2 > 90% suggests excessive recirculation and repositioning of the cannula may be necessary.

FLUIDS/NUTRITION
Fluid, electrolytes and nutritional requirements of patients on ECMO support are not substantially different from other critically ill patients. When calculating insensible losses, fluid loss through the membrane lung should be kept in mind.

The fluid loss is approximately 2 ml/m^2/hr depending on the gas flow rate across the oxygenator. In the neonate this translates into approximately 10 ml/kg/24hr.

Nutrition is supplied by total parenteral nutrition (TPN). TPN is usually infused directly into the ECMO circuit; the port immediately prior to the bladder is typically used. Enteral feeding is usually not feasible due to paralytic ileus although some groups give 1-2 ml/hr of enteral feeding for its trophic effects. Patients placed on ECMO support are usually fluid overloaded due to prior resuscitative efforts. As a consequence fluids may be restricted to 60-80 ml/kg/24hr until a steady state is realized. Diuretics are usually used to assist the fluid removal. If the patient's weight is 10% greater than dry weight, we will usually initiate a furosemide drip (0.2-0.8 mg/kg/hr). Metolazone (0.1-0.2 mg/kg PO BID) or chlorothiazide sodium (5-20 mg/kg IV BID) may be added as needed. For lesser degrees of fluid overload, intermittent furosemide (1 mg/kg given 2-4 times per day) will suffice.

For patients with gross fluid overload, > 20-25% over dry weight, or renal insufficiency, hemofiltration or dialysis may be necessary.[11] A hemofilter (Minntech, Corp., Minneapolis, MN) is placed on the venous side of the circuit with the afferent line connected to the ECMO circuit immediately after the roller pump and the efferent line connected to the bladder. Flow out of the filter is controlled by connecting the efferent line to a continuous infusion pump. If needed, dialysate can be run through the hemofilter in a direction opposite to blood flow. Nephrology consultation may be useful to determine the composition of the dialysate or flow rates. The hemofilter has a tremendous capacity to remove fluid. However, the temptation to remove the fluid quickly should be resisted. Usually a minimum of 48 hours is needed to remove a large amount of fluid. Our approach is to choose a desired goal for fluid removal for each hour. The amount chosen varies with the clinical situation but as a starting point we estimate the amount of total fluid removal desired within the next 48 hours. This is then divided by 48 to give an estimate of the hourly fluid removal that will be necessary to accomplish our goal. The entire input of the previous hour is then calculated. The desired amount of ultrafiltrate removed the following hour is equal to the total input from the previous hour plus the extra desired hourly fluid loss minus insensible losses and urine output. The amount of ultrafiltrate removed can be easily adjusted with a second continuous infusion pump attached to the line draining the ultrafiltrate from the hemofilter.

PULMONARY

Ventilator management while on ECMO support varies from program to program as there is no good data supporting one strategy over another. However, some general principles are applicable. Lung "rest" is paramount. As a consequence, the peak inspiratory pressure should be limited to approximately 20 cm H$_2$O. Alveolar recruitment is also necessary and some degree of positive end expiratory pressure should be maintained. The FiO$_2$ is usually dropped to 21-40%. If the patient is on VV bypass a slightly higher FiO$_2$ may be needed to maintain satisfactory oxygenation. Patients on VV support should also have their ventilator settings weaned slowly due to the previously described inefficiency of VV ECMO.

The lungs should be allowed to collapse only if a **severe** air leak is present. Pulmonary physiotherapy, suctioning, bronchoscopy and lavage can be used as needed although caution should be used due to anticoagulation. Typically, bilateral opacification is present on chest roentgenogram for 24-48 hours after initiation of bypass followed by gradual improvement in radiographic appearance. Surfactants may be given as the chest films improve. Some groups also prefer to change to high frequency oscillatory ventilation at this point to improve alveolar recruitment.

CARDIOVASCULAR

Hemodynamic management of patients on VV bypass is not significantly different from other critically ill patients. Patients on VV bypass will usually be maintained on moderate doses of inotropes. Dopamine hydrochloride and/or dobutamine hydrochloride are typical choices although preferences vary between institutions. Initiation of VV bypass varies in several important features from VA bypass. Since VV bypass provides no hemodynamic support, initiation of VV ECMO should be much more gradual. One critical feature to keep in mind is that initiation of VV bypass will significantly dilute any inotropes infused into the patient. To counteract this a second inotrope mixture should be started into the reinfusion portion of the circuit. Once a steady state is reached this second infusion can be slowly weaned.

Patients on VA bypass are supported hemodynamically by the ECMO circuit. Since ECMO both increases afterload and decreases preload by draining blood away from the heart a decrease in pulse pressure is virtually always noted. Some pulse pressure is desired, however, since this signifies left ventricular ejection. As noted earlier, failure to eject may lead to elevated left atrial pressures and subsequent pulmonary edema. Low grade inotropic support may assist the left ventricular ejection. However, inotropic doses should be minimized to prevent down regulation of beta-receptors. Inotropes should not be needed for hemodynamic support in patients on VA bypass. Failure to provide hemodynamic support while on VA bypass should initiate a search for other problems such as undiagnosed sepsis for example. If myocardial failure is the primary reason VA support was initiated, recovery of ventricular function can be followed by serial echocardiography. An apical short axis view can usually be obtained even with a surgical dressing in place. Pump flows should be transiently decreased to 20-25 ml/kg to volume load the ventricle while the echocardiogram is performed.

CENTRAL NERVOUS SYSTEM

Periodic neurological checks are necessary for all patients on ECMO support. As is typical in critically ill patients, full assessment usually is not feasible due to sedation. In neonates a cranial ultrasound is usually obtained every Monday, Wednesday and Friday to screen for intracranial hemorrhage. An ultrasound is also indicated if there are any changes suggesting neurological injury such as an increase in fontanelle pressure or seizures. If seizures are present phenobarbital sodium is the usual drug of choice. Electroencephalograms are used as clinically indicated.

There is no consensus regarding sedation of ECMO patients. Although some programs routinely place patients on paralytic agents we prefer to use sedation primarily and resort to muscle relaxants only when necessary for the safety of the patient. If needed a vecuronium bromide drip (0.1 mg/kg/hr) is usually initiated. Sedation is accomplished with morphine sulfate (10-50 mcg/kg/hr) and midazolam hydrochloride (0.1 mg/kg/hr). Chloral hydrate (30-50 mg/kg) may be added.

ANTIBIOTICS

4

Cefazolin sodium (25 mg/kg IV q6hr) is routinely given as prophylaxis. Since signs of sepsis such as an elevated temperature may be masked while the patient is on ECMO, routine blood, urine and sputum cultures are sent every other day.

WEANING

VENOVENOUS BYPASS

Indications of lung recovery include improved appearance on chest roentgenogram, increasing compliance and improved arterial saturation. Ventilation settings should be increased to desired settings, usually with an FiO$_2$ of 60% and moderate airway pressures. Some groups prefer to give surfactant prior to weaning. If anticoagulation has been at an ACT < 180 seconds, heparin is increased so that the ACT returns to the 180-200 second range. Flows are then gradually decreased to 20-25 ml/kg. If this is well tolerated gas flow through the membrane lung is capped off. The ECMO pump flow is kept at approximately 20-25 ml/kg. The ECMO circuit then serves in effect as a venous shunt and mixed venous oxygen saturations can be monitored continuously to assess the adequacy of systemic oxygen delivery. If the trial off is well tolerated for 4-6 hours the patient can be decannulated.

VENOARTERIAL BYPASS

Indications of myocardial recovery include an increase in pulse pressure, improved SvO$_2$ in the absence of change in other parameters and settings and improved contractility by echocardiography. Prior to initiating weaning, inotropes are infused directly into the patient. Epinephrine (0.05-0.1 mcg/kg/min) is usually our initial choice for weaning although preferences may vary between institutions. Appropriate ventilator settings are established as well. The ACT is kept at 180-200 seconds. The bypass flows are then slowly weaned. If ECMO flows of 25-30 ml/kg are tolerated a trial completely off bypass is warranted. The arterial and venous tubing leading to the patient is clamped and blood is allowed to flow through the bridge. Pump flow is maintained at approximately 100 ml/kg. The clamps and the tubing are momentarily released every 10-15 minutes to prevent clotting within the tubing. If the trial off is tolerated for 4-6 hours the cannulae are then removed.

DECANNULATION

Removal of cervically placed cannulae is performed at the bedside. Our preference is to have the assistance of surgical nurses but some programs have a decannulation kit available. The position of the patient is identical to that for cannulation. The sutures are removed and the wound irrigated with warm saline. A 2-0 silk ligature is placed around the proximal internal jugular vein. The ties around the vein are then cut with a #15 blade on the previously placed vessel loop segments. The cannula is quickly removed and a vascular clamp applied. The tie is then secured. The arterial cannula, if present, is removed in an identical manner. The wound is then irrigated, hemostasis is obtained and the wound is closed with a running 4-0 nylon suture.

Some groups prefer to repair the carotid artery, and occasionally, the internal jugular vein as well. Although this can be accomplished by well established vascular repair techniques no benefit to this repair has been demonstrated. Furthermore the incidence of thrombosis after repair appears significant and late complications such as an infected pseudoaneurysm are not altogether infrequent. Collateral flow, as demonstrated by reversal of flow in the right external carotid artery has been demonstrated shortly after cannulation. Subsequent reestablishment of flow in the common carotid artery several days later may be of questionable benefit.

Decannulation of patients with transthoracic cannulae is also performed in the ICU with the assistance of surgical nurses. Decannulation is identical to removal of cannulae following open heart surgery.

FOLLOW-UP

Protocols for follow-up of pediatric and adult patients have not been established. Concerns about long-term outcome that are paramount in neonatal patients such as cognitive development and growth are less pressing in older patients, since clinically significant neurologic impairment is usually manifest before discharge from the hospital. Long-term follow-up of the largest series of adult patients suggest that some patients have a mild restrictive pattern on pulmonary function testing but none have clinically significant lung disease.[5] Long-term follow-up should be similar to that of other critically ill patients.

Follow-up of neonatal ECMO patients is similar to other neonatal ICU unit graduates with emphasis on growth and cognitive development. The long-term follow-up protocol at The University of Iowa is shown in Table 4.6.

Table 4.6. Follow-up of neonatal extracorporeal membrane oxygenation patients

Discharge
  Medical exam
     Monitor growth and feeding
     Residual lung disease
     Incision site
  Neurodevelopmental exam
     Evaluate posture, tone, movement, primitive reflexes, sensory function, adaptive
     behavior
     Brainstem auditory-evoked response (BAER)
  Baseline head computed tomography
4-6 months
  Medical history, physical and neurological exam (as above) – repeat BAER if abnormal
  Neurodevelopmental screening (Denver II, Bayley or Gesell)
  Occupational and physical therapy evaluation and treatment if motor delay
One Year
  Medical history and physical exam (as above)
  Neurodevelopmental exam
  Behavioral audiometry
Two Year
  Medical history and physical (as above)
  Language screening (PPVT-R, Language Development Screening)
Three Year
  Medical history and physical exam (as above)
  Neurodevelopmental (formal assessment with Stanford, Binet, Gesell, McCarthy or
     IOPPSI-R)
Four Year
  Same as two year
Five Year
  Medical history, physical and neurological exam
  General intellectual functioning (a PPSI-R, McCarthy or Stanford Binet)
  Language screening (PPVT-R, verbal memory)
  Visual motor integration (Developmental Test of Visual-Motor, Integration)
  Behavior screening (Child Behavior Checklists or Conners' Behavior)
  Questionnaires, parent and teacher reports

REFERENCES

1. Bartlett RH. Extracorporeal life support for cardiopulmonary failure. Curr Probl
    Surg 1990; 27(10).
2. Field DJ, Davis C, Elbourne D et al. UK collaborative randomized trial of neona-
    tal extracorporeal membrane oxygenation. Lancet 1996; 348:75-82.
3. Shanley CJ, Hirschl RB, Schumacher RE et al. Extracorporeal life support for
    neonatal respiratory failure. Ann Surg 1994; 220:269-282.
4. Delius RE, Bove EL, Meliones JN et al. Use of extracorporeal life support in pa-
    tients with congenital heart disease. Crit Care Med 1992; 20:1216-1222.
5. Kolla S, Awad SS, Rich PB et al. Extracorporeal life support for 100 adult patients
    with severe respiratory failure. Ann Surg 1997; 226:544-566.
6. Green TP, Timmons OT, Fackler JC et al. The impact of extracorporeal mem-
    brane oxygenation on survival in pediatric patients with acute respiratory fail-
    ure. Crit Care Med 1996; 24:323-329.

7. Walters HL III, Hakimi M, Rice MD et al. Pediatric cardiac surgical ECMO: Multivariate analysis of risk factors for hospital death. Ann Thorac Surg 1995; 60:329-337.

8. Delius R, Anderson H III, Schumacher R et al. Venovenous compares favorably with venoarterial access for extracorporeal membrane oxygenation in neonatal respiratory failure. J Thorac Cardiovasc Surg 1993; 106:329-338.

9. Morris AW, Wallace CJ, Menlove RL et al. Randomized clinical trial of pressure controlled inverse ratio ventilation and extracorporeal CO_2 removed for adult respiratory distress syndrome. Am J Respir Crit Care Med 1994; 149:295-305.

10. Zapol WM, Snider MT, Hill JD et al. Extracorporeal membrane oxygenation in severe acute respiratory failure: A randomized prospective study. JAMA 1979; 242:2193-2196.

11. Heiss KF, Pettit B, Hirschl RB et al. Renal insufficiency and volume overload in neonatal ECMO managed by continuous ultrafiltration. Trans Am Soc Artif Intern Organs 1987; 10:557-560.

Ventricular Assistance for Postcardiotomy Cardiogenic Shock

Wayne E. Richenbacher

5

INTRODUCTION

Mechanical ventricular assistance is offered to patients with postcardiotomy cardiogenic shock with the expectation that ventricular recovery will occur. When the patient's ventricle has recovered function, the mechanical blood pump is removed. In general, mechanical ventricular assistance is offered to patients who cannot be weaned from cardiopulmonary bypass (CPB) despite inotropic support and intraaortic balloon counterpulsation following a technically successful open heart operation. In addition, a patient who is able to be weaned from CPB but who deteriorates hemodynamically within hours or days of his open heart procedure may also be a candidate for interim mechanical ventricular assistance. The purpose of mechanical ventricular assistance in this particular clinical application is to decompress the patient's heart thereby reducing myocardial oxygen demand. At the same time systemic perfusion is normalized. The increase in systemic perfusion improves myocardial oxygen supply and maintains end organ function. The reduction in myocardial oxygen demand and improvement in myocardial oxygen supply creates a milieu that favors ventricular recovery.

Liotta et al reported the first clinical use of a left heart assist device in 1963. In 1965, Spencer and colleagues described the first successful use of a left heart assist device in a patient who received 6 hours of left atrial-to-femoral bypass following an open heart operation. Currently, up to 6% of patients who undergo an open heart operation cannot be weaned from CPB. The majority of these patients can be managed with inotropic therapy and intraaortic balloon counterpulsation alone. Approximately 1% of postcardiotomy patients require a mechanical blood pump for management of postcardiotomy cardiogenic shock. Approximately 25% of patients who receive mechanical blood pump support for postcardiotomy cardiogenic shock survive to hospital discharge. Unfortunately, this survival rate has not improved during the past two decades. The primary determinant of survival in the postcardiotomy cardiogenic shock patient population is the degree of perioperative myocardial injury. If there is a loss of more than 40% of the myocardium ventricular function will not recover to the point where the mechanical blood pump can be removed. At this time, there is no prospective way in which to determine the amount of myocardium that has been irreversibly damaged perioperatively. Thus, if a patient has had an appropriate open heart procedure, demonstrates marginal hemodynamics despite maximum medical therapy and has no finite contraindication to mechanical ventricular assistance it is prudent to expeditiously implant a ventricular assist device (VAD) in order to provide the patient with the opportunity to recover ventricular function. Patients who require a VAD for postcardiotomy cardiogenic shock and who survive to hospital discharge have a reasonable long-term prognosis. The 2 year survival for patients who have been discharged from the hospital after having received ventricular assistance for postcardiotomy cardiogenic shock exceeds 80%.[1] More importantly, these patients enjoy a reasonable quality of life as 86% of hospital survivors are classified as New York Heart Association functional class I or II.

PATIENT SELECTION AND PREPARATION

ELECTIVE PATIENT EVALUATION

The majority of patients who require mechanical circulatory support for postcardiotomy cardiogenic shock do so on an urgent or emergent basis.[2] Although postcardiotomy ventricular dysfunction is frequently unanticipated, we make every effort to identify patients who are at risk for postoperative ventricular dysfunction prior to surgery. Patients referred for coronary revascularization are questioned regarding symptoms of congestive heart failure (Fig. 5.1). Patients who have noted a recent increase in weight, experienced dyspnea on exertion, orthopnea, paroxysmal nocturnal dyspnea or frequent nocturnal voiding require a more thorough evaluation of ventricular function. On physical examination, patients who have unexplained peripheral edema particularly toward the end of the day, basilar pulmonary rales, jugulovenous distention or a cardiac apical impulse that is displaced toward the anterior axillary line receive a similar evaluation of ventricular function. The patient's chest roentgenogram is inspected for cardiomegaly or interstitial and perivascular pulmonary edema. The latter chest roentgenogram findings occur as a result of the elevation in left atrial, pulmonary venous and pulmonary capillary pressures secondary to left ventricular dysfunction. When the pulmonary capillary wedge pressure exceeds 25 mm Hg, pleural effusions may accompany the pulmonary interstitial findings.

The cardiac catheterization is reviewed with particular attention directed toward the left ventriculogram and distribution and diffuse nature of the patient's occlusive coronary artery disease. Patients who have diminutive coronary arteries

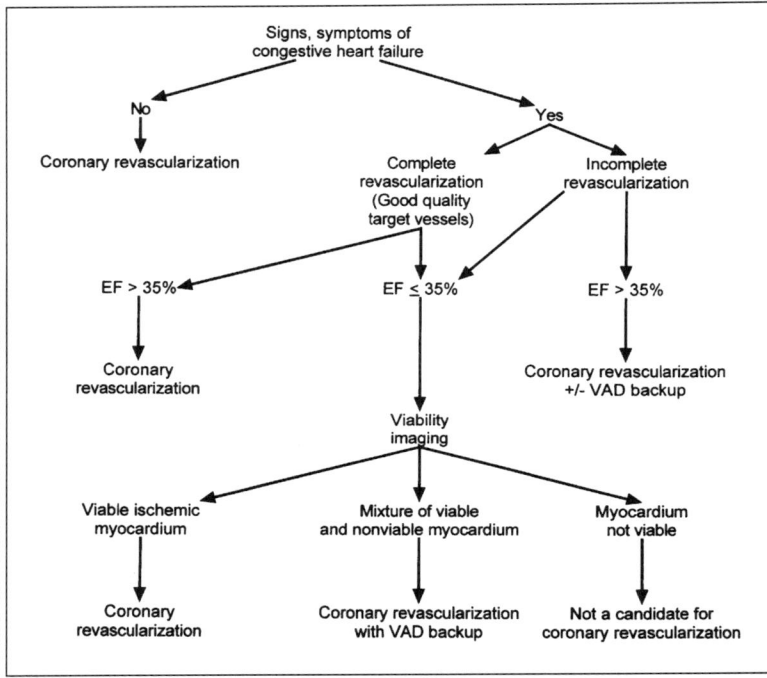

Fig. 5.1. Elective preoperative evaluation of the patient with ischemic heart disease. EF = ejection fraction. VAD = ventricular assist device.

or who may be incompletely revascularized due to diffuse atheromatous disease are at risk for postoperative ventricular dysfunction. Patients who have signs or symptoms of congestive heart failure associated with an ejection fraction below 30-35% by left ventriculography or a radioisotope ventriculogram undergo viability testing. Viability testing distinguishes viable myocardium from scar. Exercise thallium imaging documents the presence and location of ischemic myocardium (Fig. 5.2). Areas of questionable viability are further examined with positron emission tomography in order to identify metabolically active myocardium. If a patient presents with metabolically active myocardium in an ischemic wall segment supplied by a stenotic coronary artery, revascularization is often beneficial even in the presence of a depressed ejection fraction.

Patients who present with valvular heart disease undergo a similar preoperative evaluation. The right heart catheterization provides a reasonable estimate of the severity and duration of the patient's valvular heart disease. In general, the patient with normal pulmonary artery pressures will usually tolerate a valve operation even in the face of a low left ventricular ejection fraction. Conversely, valve repair or replacement in a patient with severe pulmonary hypertension, particu-

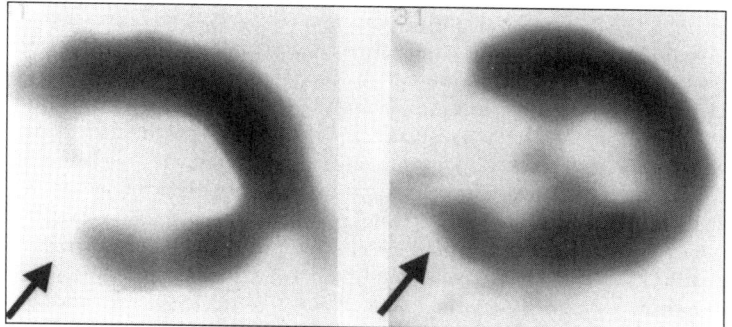

Fig. 5.2. Viability imaging. Vertical long axis views from a thallium scan (left) and [18F] deoxyglucose positron emission tomogram (FDG PET scan, right) in a patient with ischemic heart disease. The thallium scan demonstrates severely decreased thallium uptake suggesting infarct in the basilar portion of the inferior wall (arrow). The FDG PET scan demonstrates FDG uptake suggesting viable myocardium in the same region. The radiographic pattern suggests that this patient would benefit from coronary revascularization. (Courtesy of Maleah Grover-McKay, MD).

larly when the elevated pulmonary artery pressures are fixed, is a high risk undertaking. The operative risk, and hence, the potential need for mechanical circulatory support postoperatively is also determined by the valve(s) involved and the valvular condition that warrants surgical intervention. Patients with aortic stenosis and depressed left ventricular function should do well following aortic valve replacement as the obstruction to left ventricular outflow is relieved when the stenotic valve is removed. On the other hand, patients with mitral regurgitation in association with a low ejection fraction may experience an exacerbation of left ventricular dysfunction following mitral valve repair or replacement. The reduction in ejection fraction that occurs following surgical correction of mitral regurgitation is attributed to the loss of the low pressure regurgitant fraction. When the left ventricle is forced to eject blood into the high pressure systemic circulation there is a tendency for the left ventricle to dilate and decompensate.

Patients who are considered to be at high risk for postoperative left ventricular dysfunction are counseled regarding the technical limitations of surgery. The patient is informed of the possibility of ventricular dysfunction postoperatively and the potential need for an advanced form of mechanical circulatory support. The purpose of this preoperative discussion is not to dissuade the patient from undergoing operative repair but rather to provide the patient and his family with a realistic understanding of the medical problem and risks associated with the operation. A lucid preoperative discussion with the patient can potentially alleviate the need for an intraoperative, emergent family consultation regarding the need for mechanical ventricular assistance in a patient who cannot be weaned from CPB. The patient is asked to sign an operative consent form that includes a description of the primary operation as well as the possibility of mechanical blood pump insertion. Patients who are identified as being at risk for postoperative ventricular dysfunction are also subjected to an abbreviated preoperative cardiac transplant evaluation (Table 5.1). The purpose of the preoperative transplant evaluation is to facilitate intraoperative decision making regarding potential outcome. If it is clear that the patient is not a transplant candidate and the potential for ventricular recovery is small it may be inappropriate to initiate postcardiotomy mechanical ventricular assistance with the expectation that the patient can be supported with a mechanical blood pump as a bridge to cardiac transplantation should ventricular recovery not occur. If, however, the preoperative abbreviated transplant evaluation shows no obvious contraindication to cardiac transplantation the patient has yet another therapeutic option should ventricular recovery not occur following a reasonable period of mechanical circulatory support. In the latter scenario, if the patient's ventricular function does not improve over time the patient may be converted to a long-term VAD with plans to bridge the patient to cardiac transplantation. Finally, by preoperatively identifying patients at risk for postoperative ventricular dysfunction the surgical service has the opportunity to mobilize the mechanical circulatory support team to ensure that individuals specifically trained in the use of mechanical blood pumps are available at the time of the patient's open heart procedure. Having the appropriate personnel and equipment available at the time of the patient's cardiotomy greatly facilitates the conduct of the operation.

EMERGENT PATIENT EVALUATION

The careful planning and preparation that precedes elective operative intervention in a patient with ventricular dysfunction is not possible when a patient is referred for emergent surgery. The typical scenario involves a technical misadventure in the cardiac catheterization laboratory in a patient who is undergoing diagnostic coronary angiography or a catheter based intervention. Even when communication between the cardiology and cardiac surgical team is excellent the first notification usually takes place when the patient in the catheterization laboratory is in extremis. The patient is often hemodynamically unstable. Inotropes are being administered and an intraaortic balloon pump (IABP) is in place. Occasionally cardiopulmonary resuscitation is in progress. The primary determinant of survival in this instance is the time to coronary reperfusion. The patient's potential for recovery is maximized when the patient is rapidly transported to the oper-

---

Table 5.1. Abbreviated cardiac transplant evaluation

1. Age ≤ 65 years.
2. Normal pulmonary artery pressures or reversible pulmonary hypertension.
3. Normal end organ function.
4. No active infection.
5. Normal neurologic examination.
6. No malignancy.
7. Adequate financial resources (Third party payor coverage).

ating room, placed on CPB and the heart decompressed and revascularized. Emergent CPB systems that employ percutaneous peripheral cannulae can be utilized in the cardiac catheterization laboratory. In our experience, however, these systems are able to maintain systemic perfusion but cardiac decompression is incomplete and the time taken to prime the system and cannulate the patient simply delays transfer to the operating room. In our institution an operating room, perfusionist and surgical team are always available. Hence, we prefer to rapidly transport the patient the short distance from the cardiac catheterization laboratory to the operating room for a rapid sternotomy and conventional CPB.

As the patient is being transported to the surgical suite the operating surgeon should obtain a brief history including the patient's age, preprocedural functional status, comorbid conditions and baseline ventricular function. Cardiac catheterization films and details of the resuscitative effort should be reviewed. If the patient has suffered a cardiac arrest or been subjected to a prolonged period of systemic hypotension the patient's neurologic status is ascertained. These data will provide the surgeon with some sense of potential outcome and candidacy of the patient for mechanical circulatory support should the patient have refractory cardiogenic shock following coronary revascularization. If the patient is neurologically intact, had reasonable pre-event ventricular function and no irreversible comorbid conditions, it would not be unreasonable to use a VAD if the patient cannot be weaned from CPB following coronary revascularization. Even if the patient had poor ventricular function prior to the event, coronary revascularization plus VAD insertion or even primary VAD placement may stabilize the patient's condition and permit time for a more orderly assessment of the patient's potential for ventricular recovery or candidacy for cardiac transplantation. In the latter scenario, if the patient awakens following VAD insertion, has normal end organ function, does not develop an infection or recover ventricular function over the ensuing week, the patient may be returned to the operating room for insertion of a long-term VAD as a bridge to transplantation. The situation to be avoided is inappropriate VAD insertion in a patient that is clearly not salvageable or in whom there is no definable endpoint. Following coronary revascularization if a patient cannot be weaned from CPB, mechanical circulatory support would not be appropriate in a patient who experienced a prolonged period of hypotension with probable irreversible neurologic sequelae or an elderly patient with profound preprocedural ventricular dysfunction who is excluded from transplant candidacy due to age criteria.

OPERATIVE MANAGEMENT

CONDUCT OF THE OPERATION

For an elective, high risk operation the patient is typed and crossed for 2 units of packed red blood cells. The blood bank is also notified that the patient may require VAD insertion. As blood and blood product utilization at the time of VAD insertion can be excessive, the forewarning allows the blood bank time to ensure that adequate personnel and blood products are available at the time of the patient's operation should VAD insertion be necessary. Broad spectrum perioperative antibiotics provide coverage for gram positive and gram negative organisms. We utilize vancomycin hydrochloride (1 gm IV) and cefotaxime sodium (1 gm IV).

The conduct of the operation is modified to facilitate VAD insertion should that be necessary (Table 5.2). A transesophageal echocardiogram aids in evaluating wall motion and should mechanical circulatory support be necessary helps identify the presence of a patent foramen ovale. A common femoral arterial line is inserted prior to the initiation of CPB. Should the patient require IABP placement the presence of the femoral arterial line facilitates cannulation of the common femoral artery. Potential assist pump cannulation sites on the patient's heart should be preserved. In most instances CPB arterial cannulation can be accomplished in the very distal ascending aorta or transverse arch. If the ascending aorta is short or oriented transversely it may be wise to cannulate the common femoral artery. When performing coronary artery revascularization, consideration should be given to utilization of sequential vein grafts while proximal anastomoses should be oriented away from the right anterolateral aspect of the ascending aorta, the usual site for left VAD (LVAD) outlet cannula insertion. The most common site for LVAD inlet cannula insertion is the junction between the right superior pulmonary vein and the left atrium. When a patient is to undergo mitral valve replacement the standard left atrial incision through Waterston's groove may limit access to this cannulation site. Consideration should be given to using an alternate left atrial incision. Similarly, a superior transseptal incision for exposure of both mitral and tricuspid valves may make it difficult to cannulate the free wall of the right atrium in a patient who requires postoperative right ventricular assistance. Make every effort to tailor the operation to facilitate LVAD or right VAD (RVAD) insertion should mechanical circulatory support be necessary upon completion of the procedure.

The two most important considerations when performing a high risk open heart operation are to perform a technically complete operation in an expeditious fashion. A mechanical blood pump will not solve problems created by a technically incomplete operation. Furthermore, when a prolonged operation is combined with VAD insertion, CPB time is extended and accompanied by a number of adverse systemic sequelae. Activation of the cytokine and complement cascades

---

Table 5.2. Techniques that facilitate ventricular assist device insertion following an elective, high risk open heart operation

1. Obtain preoperative consent for possible ventricular assist device use.
2. Place intraaortic balloon pump leads prior to skin prep.
3. Insert common femoral arterial line prior to cardiopulmonary bypass.
4. Utilize transesophageal echocardiography.
5. Preserve assist pump cannulation sites on the patient's heart.
6. Perform a technically complete operation in an expeditious fashion.

results in a profound capillary leak syndrome. Hemodilution and trauma to formed blood elements results in a coagulopathy, and occasionally, hemoglobinuria. Postoperative bleeding can be excessive. We employ aprotinin in high risk patients in order to minimize postoperative blood loss (Table 5.3). The operative field should be hemostatic as massive blood and blood component transfusion is frequently accompanied by a rise in pulmonary artery pressures and reduction in right ventricular function. We do not utilize a blood scavenging system other than cardiotomy suction during the operation. Nor do we return blood from the mediastinal tube collection system to the patient postoperatively. In an effort to minimize CPB time the initial portion of the operation should be completed in a timely fashion and an algorithm for management of heart failure followed thereafter (Fig. 5.3).

WEANING FROM CARDIOPULMONARY BYPASS

When the open heart operation is complete, prior to terminating CPB, the vigor with which the heart is contracting can be ascertained by transesophageal echocardiography for left ventricular wall motion and visual inspection of the right ventricular free wall to determine right ventricular contractility. A vigorously contracting heart will usually wean from CPB without difficulty. However, a poorly contractile ventricle usually portends that a low output state will follow discontinuation of CPB. Maximize the potential for weaning from CPB by routine epicardial pacing in patients with a heart rate below 70-80 beats per minute. If the patient's atrioventricular node conduction is normal, atrially pace at 90 beats per minute. If a patient is in heart block employ atrioventricular pacing at a rate of 90 beats per minute with an atrioventricular delay of 150 milliseconds. If the patient's ventricular contractility appears sluggish inotropes are added empirically prior to the initial attempt at weaning the patient from CPB. If the patient appears to suffer primarily from left ventricular dysfunction we employ epinephrine (0.01-0.20 µg/kg/min IV) or dobutamine hydrochloride (2-20 µg/kg/min IV). If the patient is tachycardic or has a high systemic vascular resistance milrinone lactate (50 µg/kg IV loading dose, 0.375-0.75 µg/kg/min IV continuous infusion maintenance dose) is the drug of choice for left ventricular dysfunction. If right

Table 5.3. Aprotinin dosage for high risk patients

Test dose	1.4 mg (1 x 10^4 KIU)	intravenously
Loading dose	280 mg (2 x 10^6 KIU)	administered over 20-30 minutes beginning after induction of anesthesia, prior to sternotomy
Pump prime	280 mg (2 x 10^6 KIU)	
Constant infusion	70 mg/hr (5 x 10^5 KIU/hr)	continuous intravenous infusion until the operative procedure is completed and the patient leaves the operating room

5

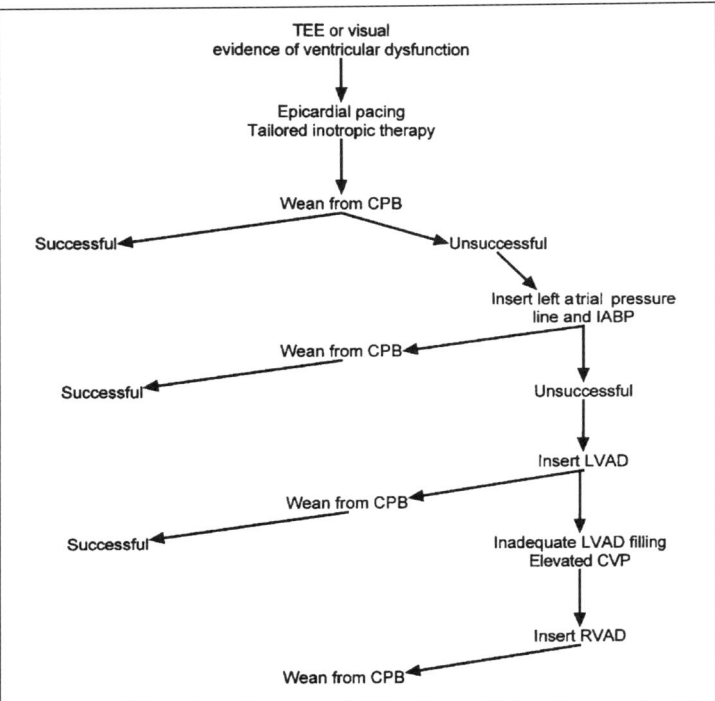

Fig. 5.3. Intraoperative management of ventricular dysfunction. TEE = transesophageal echocardiography, CPB = cardiopulmonary bypass, IABP = intraaortic balloon pump, LVAD = left ventricular assist device, CVP = central venous pressure, RVAD = right ventricular assist device.

ventricular failure is the primary problem, the ideal inotrope is isoproterenol hydrochloride (0.01-0.20 µg/kg/min IV). This inotrope improves right ventricular contractility and selectively reduces pulmonary vascular resistance facilitating right heart output. If the patient's systemic perfusion pressure permits, nitroglycerin (0.1-30 µg/kg/min IV) is added in an effort to further reduce pulmonary vascular resistance.

Do not attempt to repeatedly wean the patient from CPB with the unrealistic expectation that a patient with a marginal cardiac output will improve appreciably over time. Such a management approach invariably leads to one of two scenarios. In the first scenario the patient will be on CPB for an extended period of time and then require a VAD. Alternatively, the patient may be separated from CPB with great difficulty only to deteriorate further in the immediate postoperative period. Either situation is associated with a poor outcome. Our first attempt to wean the patient from CPB is accomplished with epicardial pacing and inotrope administration (Fig. 5.3). If it is clear that the patient has ventricular dysfunction that is refractory to inotropic therapy CPB is reinstituted (Table 5.4). A left atrial pressure line is inserted to facilitate management of left ventricular dysfunction and aid in the distinction between left and right ventricular dysfunction. The left atrial pressure line is inserted into the junction between the right superior and inferior pulmonary veins or into the right inferior pulmonary vein. The right superior pulmonary vein and the left atrial wall between the right superior and inferior pulmonary veins are preserved as a left atrial cannulation site for left ventricular assistance. An IABP is inserted via the common femoral artery. The second attempt to wean the patient from CPB is accomplished with inotrope administration and 1:1 intraaortic balloon counterpulsation. Immediately upon termination of CPB a thermodilution cardiac output is obtained. The majority of patients in cardiogenic shock following an open heart operation manifest left ventricular dysfunction either alone or in combination with right ventricular failure. If the patient fulfills the criteria listed in Table 5.4, CPB is reinstituted and preparations made to insert an LVAD.

INITIATION OF MECHANICAL VENTRICULAR ASSISTANCE

Temporary left ventricular assistance is accomplished using left atrial inflow and ascending aortic outflow. Left ventricular apex inflow cannulation is reserved for patients who are being bridged to cardiac transplantation. When ventricular recovery occurs, left atrial inflow cannulation simplifies decannulation and LVAD removal. Currently, two devices are approved by the Food and Drug Administration (FDA) for temporary ventricular support in patients with postcardiotomy cardiogenic shock. Both the Abiomed BVS 5000 pneumatically driven pulsatile VAD (Abiomed, Inc., Danvers, MA) and the Thoratec pneumatically driven pulsatile VAD (Thoratec Laboratories Corp., Pleasanton, CA) can be utilized for left or right ventricular assistance.[3,4] Two pumps can be employed for biventricular assistance. Although not approved for use as a long-term circulatory support device, the ubiquitous centrifugal pump can be readily adapted for use as a VAD by clinicians who do not have access to a pulsatile blood pump. For left ventricular assistance using left atrial cannulation, the inlet and outlet cannulae are brought through the soft tissues and skin of the anterior abdominal wall in the right subcostal region (Fig. 5.4). The inlet cannula is exteriorized in the lateral position.

Table 5.4. Definition of left ventricular failure

Profound left ventricular wall motion abnormalities by transesophageal echocardiography.	
Left atrial pressure	≥ 15-20 mm Hg
Peak systolic aortic pressure	≤ 90 mm Hg
Cardiac output index	≤ 2.0 L/min/m²
Despite: Functional cardiac rhythm	
Inotrope administration	
Intraaortic balloon counterpulsation	

Fig. 5.4. Left atrial-to-aortic left ventricular assistance for postcardiotomy
cardiogenic shock.

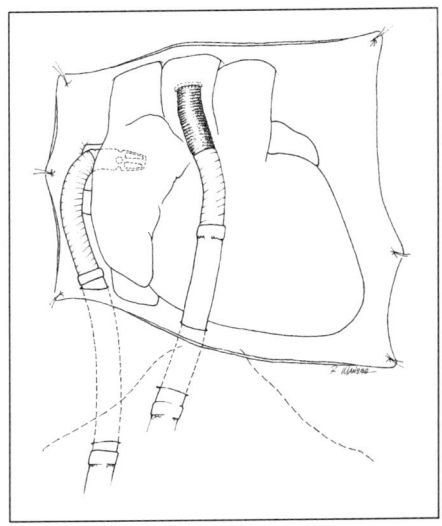

The outlet cannula is located in the medial position. Both cannulae are exteriorized to the right of the midline. The technique we use in creating the cannula
tunnels begins with the skin button. Inlet and outlet cannula skin sites are marked
2 fingerbreadths below the right costal margin. The skin is elevated with an Allis
tissue forcep (V. Mueller, Deerfield, IL) and a circular, full thickness button of skin
no greater than 1 cm in diameter is excised immediately beneath the clamp
(Fig. 5.5). The underlying fascia is exposed and a cruciate fascial incision created
with electrocautery. The goal in creating the subcutaneous tunnels is to ensure
that the fascia and skin lie snugly against the inlet and outlet cannulae. Close apposition of fascia, skin and the cannulae minimizes the potential for egress of
pericardial fluid along the cannulae tunnels and in the case of a velour covered
cannula promotes tissue ingrowth. Left atrial inflow cannulation is possible via
the left atrial appendage, dome of the left atrium or left atrial free wall between
the right superior and inferior pulmonary veins (Fig. 5.6). Left atrial cannulation
is usually accomplished by inserting the inflow cannula through the left atrial wall
near the junction of the right superior pulmonary vein and left atrium (Fig. 5.7).
With the patient on CPB the surgical assistant retracts the free wall of the right
atrium toward the patient's left. Develop the plane between the atria in order to
create a larger surface area of thickened atrial wall. Two concentric 2-0 Prolene
(Ethicon, Inc., Somerville, NJ) pursestring sutures are placed at that location. Felt
pledgets (3 x 7 mm, prepunched TFE polymer pledgets; Davis and Geck, Danbury,
CT) are included at each external passage of the needle on both pursestrings. The
sutures are passed through Rumel tourniquets constructed of a 15 cm length of a
20 F red rubber catheter. The inlet cannula is passed through the subcutaneous
tunnel prior to inserting the tip into the left atrium. If the cannula tip is inserted

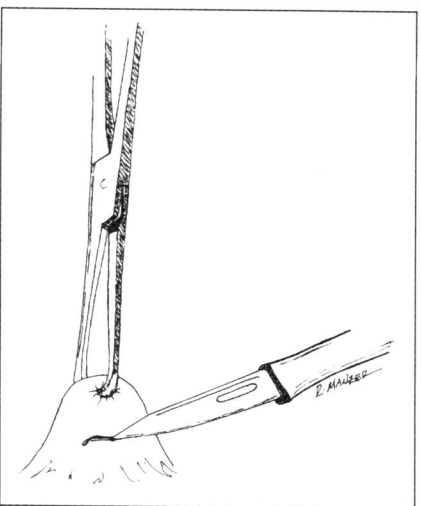

Fig. 5.5. Technique for creation of the skin exit site for blood pump cannulae. The button of skin removed should be quite small (< 1 cm diameter) to create a snug interface between the skin and cannula.

5

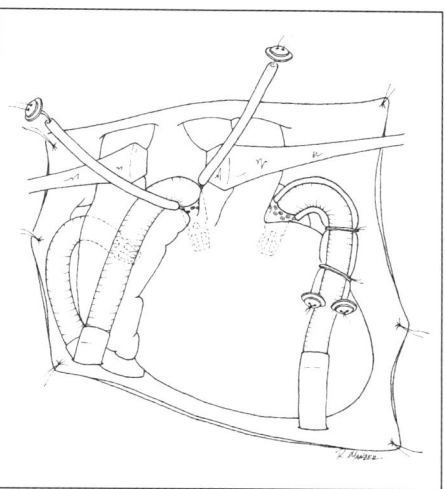

Fig. 5.6. Left atrial inflow cannulation sites for left ventricular assistance include the atrial appendage, dome of left atrium (accessed between the ascending aorta and superior vena cava) and the interatrial groove between the right superior and inferior pulmonary veins.

into the atrium first, undue tension can be placed on the entry point when the cannula is passed through the subcutaneous tunnel. The additional manipulation may contribute to postoperative bleeding around the atrial cannulation site. CPB flow should be reduced and positive pressure placed on the endotracheal tube at the time of atrial cannulation. The elevation in left atrial pressure reduces the potential for air embolism. The left atrial cannulation site is incised with a scalpel

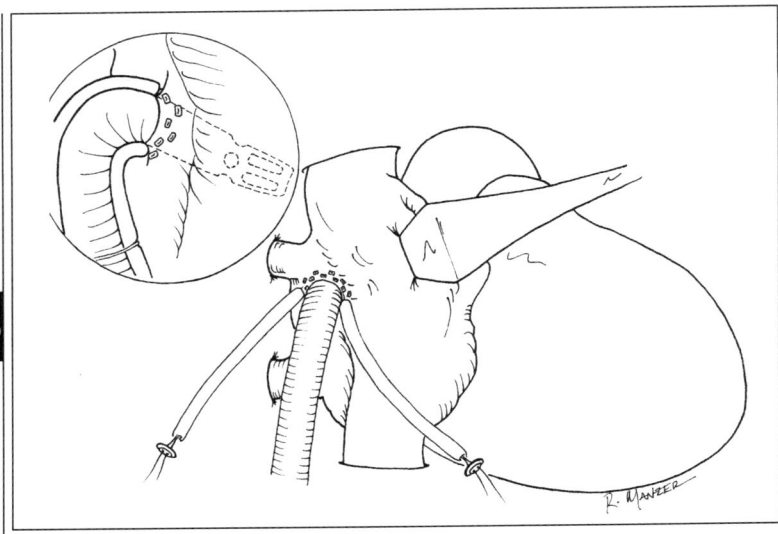

Fig. 5.7. The most common site for left atrial inflow cannulation is the wall of the left atrium between the right superior and inferior pulmonary veins. Two buttressed concentric pursestring sutures are placed in the left atrial wall (inset). After the cannula is inserted the Rumel tourniquets are snared, tied in place with polypropylene buttons and fixed to the cannula.

blade and the inlet cannula inserted through the pursestring sutures. Brief cardiac compression allows this cannula to be filled with blood and deaired. A tubing clamp is applied to the nonwire reinforced segment of the cannula along the patient's anterior abdominal wall. The pursestring sutures are snared and the suture ends tied over 14 mm polypropylene buttons (Ethicon, Inc., Somerville, NJ). The tourniquets are fixed to the cannula in two locations with braided sutures. At the time of insertion the angled tip of the inlet cannula should be oriented toward the central orifice of the mitral valve. If the tip of the atrial cannula is positioned within the body of the left atrium, flow of blood into the inlet cannula is enhanced. The tip of the inlet cannula should not be permitted to cross the mitral valve. If the inlet cannula transverses the mitral valve orifice it will be difficult to document the return of ventricular function. When the patient recovers ventricular function and LVAD flows are reduced, mitral regurgitation created by a malpositioned inlet cannula precludes ventricular ejection into the ascending aorta.

The graft portion of the outlet cannula is cut to the appropriate length and anastomosed end-to-side to the right anterolateral aspect of the ascending aorta using a continuous 4-0 polypropylene suture technique. Unless aortic length or exposure is a problem the anastomosis between the outlet graft and ascending aorta is accomplished with a partial occlusion clamp (Fig. 5.8). Upon completion of the outlet graft-to-aortic anastomosis, the partial occlusion clamp is removed

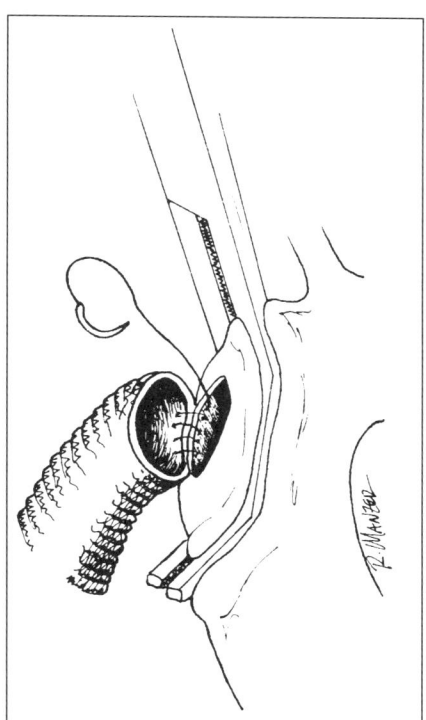

Fig. 5.8. The end of the left ventricular assist device outflow graft is sutured to the side of the ascending aorta. If aortic length permits, this anastomosis is accomplished with a partial occlusion clamp. Otherwise an aortic crossclamp is applied and the anastomosis performed during a brief period of cardioplegic arrest.

from the ascending aorta and a vascular clamp applied to the outlet graft. This allows the distal anastomosis to be checked for hemostasis. The outlet conduit is passed through the subcutaneous tunnel following completion of the graft-to-aortic anastomosis. The outlet cannula is exteriorized through the medial skin button. For centrifugal pump ventricular assistance, standard CPB cannulae are employed for VAD inflow and outflow. The outlet cannula is inserted into the ascending aorta through pursestring sutures. The blood pump and cannulae are carefully deaired as the VAD is connected to the inlet and outlet cannulae. VAD preparation and device specific implantation techniques are reviewed in Chapter 11: Device Specific Considerations.

LVAD pumping is initiated at a minimum fixed rate or low flow rate in the case of centrifugal pump left heart support. CPB flow is serially decreased as LVAD flow is increased. LVAD flow is raised as permitted by pump filling. As CPB is discontinued pay close attention to the left atrial and central venous pressures (Table 5.5). If the inflow conduit is positioned satisfactorily the left atrial pressure should remain below 5 mm Hg as LVAD flow is increased. Low left atrial pressure, in the face of satisfactory LVAD flow, indicates that there is adequate decompression of the left atrium with a brisk flow of blood into the LVAD inlet cannula. A

Table 5.5. Hemodynamic status during mechanical left ventricular assistance

CVP (mm Hg)	LAP (mm Hg)	Systolic AoP (mm Hg)	CI (L/min/m²)	Diagnosis
< 20	< 10	> 90	> 2.0	Satisfactory pumping
< 15	< 10	< 90	< 2.0	Hypovolemia
< 20	> 20	< 90	< 2.0	Inlet cannula obstruction
> 20	< 10	< 90	< 2.0	Right ventricular failure

CVP = central venous pressure
LAP = left atrial pressure
AoP = aortic pressure
CI = cardiac output index

malpositioned left atrial cannula results in inlet cannula obstruction. Inlet cannula obstruction is characterized by an elevated left atrial pressure with inadequate LVAD flow. Transesophageal echocardiography is an invaluable aid in documenting satisfactory left atrial decompression and proper positioning of the tip of the left atrial cannula. Under ideal circumstances, the left atrium and left ventricle are completely decompressed with little or no ejection from the native left ventricle. If left sided drainage is complete there is no ejection from the native left ventricle and the patient's aortic valve will not open. This state is rarely achieved in a postcardiotomy cardiogenic shock patient supported with an LVAD using left atrial cannulation.

As CPB is discontinued and the LVAD flow rate is increased, it is also important to observe the central venous pressure. If LVAD flow is suboptimal and the left atrial pressure is low, the patient may be suffering from either hypovolemia or right heart failure. Hypovolemia is readily identified by a low central venous pressure. Judicious volume administration will raise the central venous and left atrial pressure, improve LVAD filling and flow. On the other hand, the initiation of left ventricular assistance may unmask right ventricular failure (Table 5.6). Suboptimal LVAD flow in the face of a low left atrial pressure and high central venous pressure is pathognomic of right ventricular failure.

If the central venous pressure exceeds 15-20 mm Hg we initially treat right ventricular dysfunction with isoproterenol hydrochloride (0.01-0.20 μg/kg/min IV). Isoproterenol hydrochloride selectively lowers pulmonary vascular resistance and improves right ventricular contractility. The utility of isoproterenol hydrochloride is limited in patients who have ventricular tachyarrhythmias. If the systemic blood pressure permits we also administer intravenous nitroglycerin (0.1-30 μg/kg/min IV) in an effort to lower pulmonary artery pressures and improve flow through the right heart. Although not currently approved by the FDA for clinical use, the third drug of choice is inhaled nitric oxide (20-60 parts per million). Nitric oxide selectively lowers pulmonary vascular resistance. If nitric oxide is unavailable, selective pulmonary vasodilation can be achieved with intravenous prostaglandin E_1 (30-150 ng/kg/min).[5] A problem with prostaglandin E_1

Table 5.6. Definition of right ventricular failure

Cardiac output index	≤ 2.0 L/min/m^2
Right atrial pressure	≥ 20 mm Hg
Left atrial pressure	≤ 10 mm Hg

Distended right ventricle
Hypocontractile right ventricular free wall
Inability to volume load the left ventricle

use is that this drug is incompletely metabolized in one pass through the lungs.
Thus, when prostaglandin E$_1$ enters the systemic circulation the systemic vascular
resistance may fall creating dangerously low systemic arterial blood pressure. The
low systemic vascular resistance can be counteracted by administering norepinephrine bitartrate (0.03-0.06 µg/kg/min IV) directly into the left atrium. Administration of prostaglandin E$_1$ through a central venous line effectively reduces pulmonary vascular resistance while systemic vascular resistance is maintained by
administering norepinephrine bitartrate directly into the left side of the heart. As
the left atrial pressure can no longer be monitored with this particular drug regimen, a second left atrial pressure line is usually required. Nitric oxide has eliminated the need for prostaglandin E$_1$. Rather than resort to multiple drug infusions
into both sides of the heart, if the patient's right ventricle remains distended with
an elevated central venous pressure and inadequate LVAD filling despite isoproterenol hydrochloride and nitric oxide administration it is best to proceed directly
to RVAD implantation. As discussed above, the centrifugal pump, Abiomed and
Thoratec VADs, can be configured for use as either a left or right ventricular assist
pump. For right ventricular assistance inflow is achieved by inserting the RVAD
inlet cannula through pursestring sutures placed in the free wall of the right atrium.
The RVAD outlet graft is sewn in an end-to-side fashion or inserted, as in the case
of the centrifugal pump, to the proximal portion of the main pulmonary artery.
Cannulation techniques for right ventricular assistance are identical to those described for left ventricular assistance. For right ventricular assistance the cannulae
are exteriorized in the left subcostal region. The inlet cannula is positioned medially while the outlet cannula is located laterally (Fig. 5.9). The RVAD inlet cannula
and the LVAD outlet cannula must cross in the midline. Care should be taken to
ensure that the flexible, nonreinforced segment of the Abiomed or Thoratec LVAD
outlet cannula is not compressed by the rigid RVAD inlet cannula. Similarly, cardiac or vein graft compression may be problematic. The latter situations may only
become apparent at the time of sternal approximation. The cannulae and RVAD
are deaired and the device is connected to the drive unit.

 LVAD pumping is initiated followed by RVAD pumping as CPB flow is simul-
taneously reduced. As CPB is discontinued, closely observe the left atrial pressure.
With biventricular support it is possible to raise RVAD flow above the level of
LVAD flow. Should such a scenario occur, a dangerous situation results in which

Fig. 5.9. Cannulation scheme for biventricular assistance for postcardiotomy cardiogenic shock. Left ventricular assistance employs left atrial inflow and aortic outflow cannulae. Right ventricular assistance utilizes right atrial inflow and main pulmonary artery outflow cannulae.

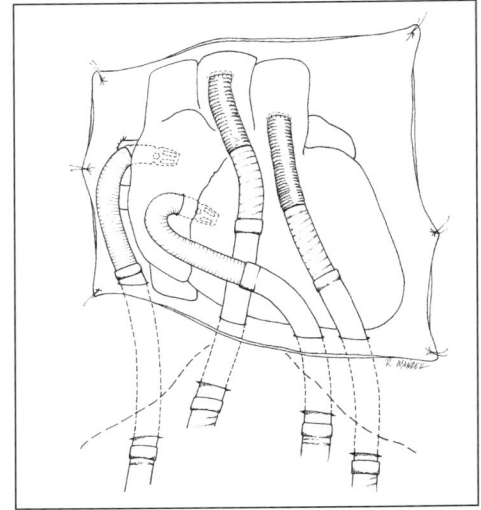

5

the pulmonary vasculature is overperfused. Similarly, should there be an impediment to the flow of blood into the LVAD inlet conduit, pulmonary hyperperfusion may occur as RVAD flow is increased. The desired endpoint is an LVAD flow index greater than 2.0 L/min/m^2 with a left atrial pressure less than 5-10 mm Hg. Regardless of whether left ventricular assistance or biventricular assistance is employed it is desirable to reduce or discontinue as many cardiotonic drugs as possible. By minimizing inotrope administration the myocardium is most effectively rested thereby facilitating the recovery of ventricular function. In addition, the elimination of alpha agents improves end organ perfusion. The sole exclusion to this rule is the use of low dose dopamine hydrochloride (3 µg/kg/min IV) to enhance renal perfusion.

Following the initiation of left ventricular assistance, discontinuation of CPB and prior to removing the CPB cannulae, transesophageal echocardiography is performed. The purpose of the transesophageal echocardiogram is to ensure that the left heart is decompressed and there is a brisk flow of blood into the LVAD inlet cannula. Secondly, a bubble study is performed to determine if the foramen ovale is patent (Fig. 8.3). In the LVAD patient even a mild rise in central venous pressure can result in arterial desaturation as unoxygenated blood passes from the distended right atrium to the decompressed left atrium by way of the patent foramen ovale. The bubble study is performed by filling a 10 ml syringe with 9 ml of saline and 1 ml of air. The syringe is vigorously shaken by hand. Any gross air within the barrel of the syringe is expelled and the remaining microbubble solution is injected rapidly into the central venous catheter. The atria are observed by echocardiography as the solution is injected. A patent foramen ovale is present when microbubbles traverse the atrial septum and appear in the left atrium. In the patient supported with both an RVAD and LVAD, the sensitivity of the bubble study can be increased by momentarily decreasing RVAD flow as the bubble study is performed. The temporary rise in central venous pressure and fall in left atrial pressure will usually force a few bubbles across a patent foramen ovale.

If the foramen ovale is patent it must be closed at the initial operation regardless of whether or not arterial desaturation has occurred. If the foramen ovale is left open the patient may suffer acute arterial desaturation postoperatively. Even a brief period of arterial desaturation can result in a profound neurologic deficit. In order to close the patent foramen ovale the right atrium must be opened. The two-stage venous cannula used for CPB is converted to two single venous cannulae which are advanced into the superior and inferior vena cavae, respectively. CPB is briefly reinstituted and mechanical ventricular assistance discontinued. The VAD cannulae are clamped to avoid the passage of air into the VAD. Caval snares are applied, a right atriotomy performed and the atrial septum exposed. The foramen ovale can usually be closed with a running polypropylene suture. Larger defects may require patch closure. CPB is again discontinued as VAD flow is resumed.

TERMINATION OF THE OPERATION

When the patient has achieved hemodynamic stability on mechanical circulatory support the CPB cannulae are removed. Protamine sulfate is administered in a dose sufficient to allow the activated clotting time (ACT) to return to the baseline level. Patients who require mechanical circulatory support for postcardiotomy cardiogenic shock experience a prolonged CPB time with its attendant coagulopathy. When protamine sulfate administration is complete a blood sample should be sent for a prothrombin time, partial thromboplastin time, platelet count, fibrinogen level and thrombin time determination. Blood and blood product administration are directed by the results of these laboratory studies.

The mediastinum is drained with large diameter tubes. If a patient requires biventricular support it may take some creativity to identify sites at which the chest tubes can traverse the soft tissues of the patient's upper abdomen. The mediastinum is frequently crowded with inlet and outlet cannulae. This is further complicated by the presence of coronary artery bypass grafts. To reduce postoperative blood loss, attempt to close the sternum. Graft and cardiac compression by the VAD cannulae can often be avoided by opening both pleural spaces thereby allowing the cannulae to lay lateral to the heart. The sternum is wired closed and VAD flow(s) observed for several minutes prior to completing the soft tissue closure. If sternal closure results in a significant reduction in VAD flow the sternum should be stented open. If the sternum is stented open the sternal wound should be covered with an occlusive dressing to reduce the potential for mediastinal sepsis.

Following standard closure of the sternotomy a sterile dressing is applied to the incision. VAD cannulae exit sites in the upper abdomen are lightly coated with 1% silver sulfadiazine cream. As percutaneous cannulae dressing changes are performed frequently it is helpful to employ Montgomery straps (Dermicel; Johnson & Johnson Medical, Inc., Arlington, TX) to hold the cannulae dressing in place.

The Montgomery straps eliminate the need for repeated application of adhesive tape to the skin on the patient's upper abdomen.

POSTOPERATIVE MANAGEMENT

The most common early postoperative problem is bleeding (Table 5.7). The mediastinal tubes should be stripped to ensure patency. The patient's hemodynamics are closely observed for evidence of cardiac tamponade. In addition to the classic signs, a more subtle indication of cardiac tamponade may be atrial compression with inlet cannula obstruction accompanied by a progressive decline in VAD flow. The coagulopathy identified by laboratory studies is readily corrected by the administration of protamine sulfate and fresh frozen plasma. A low fibrinogen level or release of fibrin split products indicative of fibrinolysis is most effectively treated with cryoprecipitate. The qualitative platelet defect associated with CPB is infrequently accompanied by a fall in platelet count. Thus, platelet pack administration may be necessary even in the face of a platelet count above 100 k/mm^3.

Regardless of the VAD type employed, postoperative anticoagulation is required in all patients. Anticoagulation is not begun until postoperative bleeding is controlled. Systemic anticoagulation is not initiated until mediastinal tube drainage is less than 50 ml/hr and is rarely started prior to 12-24 hours following VAD insertion. At that time, heparin sodium is administered by continuous intrave-

---

Table 5.7. Principles of postoperative management

Anticoagulation
    Complete heparin sodium reversal with protamine sulfate.
    Blood component therapy directed by specific coagulation defect.
    Time to initiation of postoperative anticoagulation determined by mediastinal tube drainage.
Fluid balance
    Minimize intake
        Concentrate drips
        No maintenance intravenous fluids
    Maximize output
        Continuous furosemide infusion
Infection control
    Perioperative antibiotics
    Care of percutaneous cannulae sites
    Avoid nosocomial infections
        Early extubation
        Frequent line changes
        Discontinue urinary drainage catheter
Nutritional repletion
    Early tube feedings
    Caloric supplements
Physical rehabilitation

nous infusion; the endpoint is an ACT of 180-200 seconds. A bolus of heparin sodium at the initiation of the intravenous infusion is not recommended. A platelet count determination is performed daily in patients receiving heparin sodium for an extended period of time to allow early recognition of heparin associated thrombocytopenia. As the duration of mechanical blood pump support for postcardiotomy cardiogenic shock is brief, warfarin sodium administration is not necessary.

Particular attention should be directed toward fluid balance. The systemic inflammatory response associated with CPB creates a capillary leak syndrome that is manifest by interstitial edema. It is imperative that fluid intake be minimized while output is maximized. To achieve this goal all intravenous infusions are maximally concentrated. No maintenance intravenous fluids are provided. A forced diuresis is accomplished with a continuous furosemide infusion (10 mg/100 ml normal saline, titrate intravenous infusion rate to achieve a urine output 100-200 ml greater than input each hour). Bolus furosemide administration leads to a precipitous fall in cardiac filling pressures and a reduction in VAD flow. Continuous furosemide infusion, on the other hand, allows a steady ongoing diuresis without the attendant fall in filling pressures. Accurate intake and output records are maintained. Output should always exceed input during the first 24-48 hours. Diuresis continues until the patient's weight falls below its baseline level or the patient develops significant prerenal azotemia.

Avoidance of infection is key to a successful outcome. Infection prophylaxis begins with gram positive and gram negative perioperative antibiotic coverage as previously described. Antibiotics are discontinued when the mediastinal drainage tubes are removed. Antibiotics are unnecessary beyond the first 24-48 hours. In fact, continuation of antibiotics beyond this time point is often counterproductive in that it can lead to an overgrowth of resistant organisms. With a few exceptions (a patient with an open sternum) aggressive diuresis usually permits early extubation and avoidance of nosocomial pneumonia. The metabolic alkalosis associated with the continuous furosemide infusion must be corrected prior to extubation in order to avoid postextubation hypercarbia. Shortly following the initiation of furosemide therapy acetazolamide (250-500 mg IV q6h) is added to the drug regimen. Once mechanical ventricular assistance is initiated the IABP is superfluous. Following correction of the immediate postoperative coagulopathy the IABP is removed. This will allow the patient to sit up following extubation. The change from the supine to the sitting position facilitates pulmonary toilet. Chronic obstructive pulmonary disease is aggressively treated with nebulizer treatments. Formal chest percussion should be used with care, if at all, due to the potential for dislodgement of the VAD cannulae. Invasive hemodynamic monitoring lines are removed as soon as possible. When it is clear that the hemodynamics have stabilized the left atrial pressure line is removed. Mediastinal tube removal follows withdrawal of the left atrial pressure line. If lines must remain in place the intravenous/intra-arterial catheters should be moved to new sites every 72 hours. Percutaneous cannulae sites have sterile dressing changes performed every 8 hours. Our routine involves washing the skin site with hydrogen peroxide (3% diluted 50:50

with normal saline) and covering the site with 1% silver sulfadiazine cream and a sterile dressing. As described above, the use of Montgomery straps eliminates the need for repeated application of adhesive tape to the skin on the patient's upper abdomen. The midline skin incision is left open to air beginning 24 hours following the patient's open heart operation.

Nutritional repletion begins as soon as stable hemodynamics have been achieved and the IABP removed. Provided the patient's postoperative ileus has resolved, a soft silastic feeding tube is inserted and tube feeds instituted. Nutritional repletion is best accomplished with enteral feedings as central intravenous hyperalimentation is associated with a finite infection rate. Furthermore, enteral feedings bathe the gastric mucosa decreasing the risk of stress ulceration and the potential for bacterial translocation in the gut. If intestinal motility is sluggish in the immediate postoperative period, feedings should begin with an elemental diet. When peristalsis has returned a less expensive balanced formula is employed. The endpoint is full caloric intake via tube feeds within 24-48 hours of VAD insertion. Following extubation the patient is offered clear liquids and advanced to a regular, no added salt diet as tolerated. Avoidance of salt in the diet in the early postoperative period reduces problems with fluid retention and interstitial edema. By continuing tube feeds the patient is permitted to resume a diet at a comfortable rate while oral intake is not forced. Dietary supplements in the form of high caloric milk shakes are a routine. When the patient's appetite improves the tube feedings can be cycled to nighttime only. Calorie counts are monitored, and when the full caloric requirement is met by the oral route enteral feedings are discontinued.

Rehabilitation is aggressive and begins with range of motion exercises within 24 hours of VAD insertion. The patients are allowed to stand at the bedside, ambulate and utilize a bed or stationary bicycle at an appropriate time as determined by the physical therapists. The Abiomed VAD is gravity filled and thus the activity level may be limited compared to the paracorporeal Thoratec VAD.

Diagnostic studies are dictated by the patient's clinical condition. In general, a complete blood count, serum electrolyte determination and liver function panel are obtained daily. The white blood cell count should include a differential in an effort to provide early identification of septic complications. A daily plasma hemoglobin documents the level of red blood cell destruction related to a prolonged CPB time or early cell death following a massive packed red blood cell transfusion. Hemolysis is usually unrelated to the VAD as pneumatic, pulsatile pumps are able to pump blood gently without ongoing injury to formed blood elements. If the patient has normal renal function the plasma hemoglobin level usually returns to normal (< 5 mg/dl) within 48-72 hours. Serum creatine kinase (CK) levels with MB band determination and troponin I levels document the presence and severity of a perioperative myocardial injury. A serum prealbumin level is measured once or twice weekly to determine the patient's nutritional status. Chest roentgenograms are obtained daily with particular attention directed to cannulae position, size of the cardiac silhouette and clarity of lung fields. If the cannulae are positioned correctly and the heart adequately decompressed, the cardiac silhouette should be normal in size. With forced diuresis the postoperative pulmonary interstitial edema usually clears within 48-72 hours of VAD insertion. If a patient develops a fever, defined as a temperature $\geq$ 38.5° C, an attempt is made to identify the source of sepsis. Blood cultures are drawn through two separate peripheral sites using a povidone-iodine skin prep. Sputum is obtained and sent for both gram stain and sputum culture. Urine is sent for microscopic urinalysis and urine culture. Empiric, broad spectrum antibiotic coverage is only initiated if the patient develops persistent fever associated with leukocytosis. Antibiotic coverage is tailored pending the results of body fluid cultures.

WEANING FROM MECHANICAL CIRCULATORY SUPPORT

5

Although patients have recovered ventricular function weeks or months following initiation of mechanical ventricular assistance, the usual time course of left ventricular recovery is 7-10 days. Right ventricular recovery occurs more quickly. There are two schools of thought regarding the appropriate time for VAD removal. As the presence of the mechanical blood pump represents an ongoing source of infection, thromboembolism and impairs the ability of the patient to rehabilitate, some clinicians prefer to wean the patient from the mechanical blood pump in the early postoperative period. Thereafter, the patient is supported with inotropes and possibly intraaortic balloon counterpulsation. Proponents of this management scheme argue that early VAD removal reduces the potential for development of a blood pump related complication. The second school of thought favors a more prolonged period of mechanical circulatory support. As systemic perfusion is maintained, end organ function is optimized and the myocardium is afforded the opportunity for complete recovery. If this therapeutic approach is followed the mechanical blood pump can usually be removed without the need for subsequent inotropic support. Our preference is the latter management approach as the scenario in which a patient with incomplete myocardial recovery requires inotropic support with the attendant adverse effect upon the recuperating myocardium and end organ function is avoided. The patient who demonstrates complete ventricular recovery during the period of mechanical circulatory support is much easier to manage following VAD removal.

The timing of VAD removal is determined by the patient's ventricular performance on ever decreasing levels of VAD support. In a patient supported with an LVAD and in whom an arterial line is present the arterial pressure trace can be examined for evidence of ejection from the native left ventricle (Fig. 5.10). LVAD flow is reduced by 1-2 L/min allowing the native left ventricle to fill and eject if ventricular recovery has occurred. If a Swan-Ganz catheter is present during this period of reduced LVAD support, a cardiac output is determined. Subtracting the LVAD flow from the thermodilution cardiac output, the contribution of native ejection to the total cardiac output is determined. If the patient has had an uneventful hospital course following LVAD insertion, however, invasive hemodynamic monitoring lines have usually been removed. An alternate method of monitoring left ventricular recovery is to perform transthoracic echocardiography while

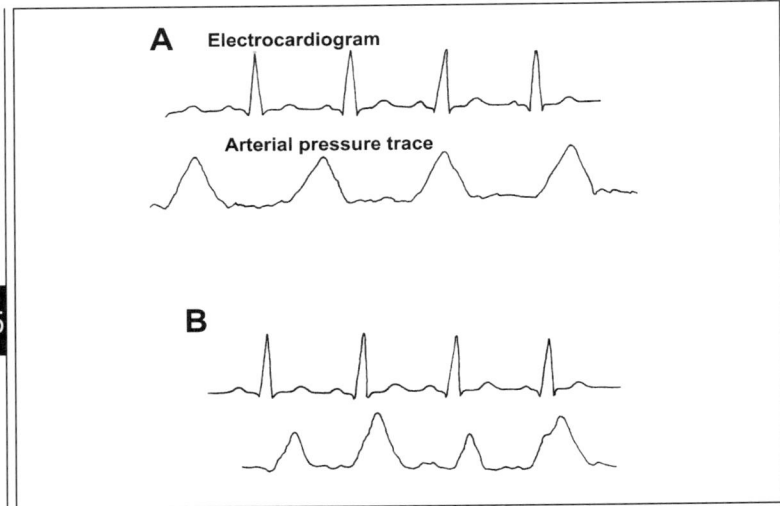

Fig. 5.10. As a patient recovers ventricular function, a reduction in ventricular assist device flow allows the native left ventricle to fill and eject. A. The arterial pressure trace is generated entirely by the mechanical blood pump. There is no association between the electrocardiogram and arterial pulsation. B. The patient has now recovered ventricular function. As ventricular assist device flow is reduced, the native ventricle fills and ejects with each electrical depolarization.

LVAD flow is serially reduced. As LVAD flow decreases, the echocardiogram in a patient in whom ventricular recovery has occurred demonstrates concentric contraction of the left ventricular free wall and septum with ejection through the native aortic valve. Conversely, patients in whom ventricular recovery is incomplete will develop ventricular distention with limited motion of the left ventricular free wall and little or no ejection through the native aortic valve. The return of ventricular function is more definitively documented with radionuclide ventriculography or transesophageal echocardiography. We find these more cumbersome, invasive, time consuming diagnostic modalities to be unnecessary. Regardless of monitoring modality employed, it is important to avoid a prolonged period of decreased LVAD flow as stasis within the mechanical blood pump increases the risk of thromboembolism. In our opinion, LVAD flow should never be decreased below 2-3 L/min.

Patients with biventricular dysfunction who require both an LVAD and RVAD are more difficult to wean from mechanical circulatory support. At no time should LVAD support be reduced without simultaneously decreasing RVAD flow. Pulmonary congestion can occur if LVAD flow is decreased without reducing RVAD flow. In the absence of left ventricular recovery, if LVAD flow is decreased, ongoing right ventricular support can create life threatening pulmonary edema. The usual approach is to serially decrease the level of both left and right ventricular assistance while observing the hemodynamic parameters and echocardiographic findings described above. It could be argued that acutely volume loading a dysfunctional ventricle will not provide a meaningful estimate of the ventricle's contractile reserve. Thus, it has been proposed that weaning schedules be developed and the patient be supported at a reduced level of ventricular assistance for a more prolonged period of time. If the patient is able to maintain systemic or pulmonic perfusion in the face of reduced left or right heart support for hours, rather than minutes, the clinician may be more comfortable in returning that patient to the operating room for device removal. Although this is certainly true the relative risk of stasis induced thromboembolism must be considered. Our routine includes a brief daily trial of reduced VAD flow beginning on post implant day three or four. The patient's hemodynamic response to this brief period of reduced VAD flow provides a reasonable impression of the degree to which ventricular recovery has occurred. Prior to returning the patient to the operating room for device removal, we reduce the VAD flow 0.5 L/min every 2-3 hours for 6-12 hours prior to VAD explantation. VAD flow is never decreased below 2 L/min during this formal weaning trial and an ACT is determined every 2 hours to ensure the level never falls below 180 seconds.

Occasionally, in the patient supported with both an RVAD and LVAD, right ventricular recovery precedes left ventricular recovery. If by post-VAD implantation day three or four RVAD flow can be decreased to 3 L/min and LVAD filling is not impaired the patient may be returned to the operating room for RVAD removal and mediastinal irrigation. The chest is closed leaving the LVAD in place and the patient returned to the intensive care for additional convalescence. When left ventricular recovery occurs 4-5 days later, the patient is returned to the operating room for LVAD removal.

DEVICE REMOVAL

A type and cross is performed for 2 units of packed red blood cells. The patient receives preoperative antibiotic coverage for both gram positive and gram negative organisms (vancomycin hydrochloride 1 gm IV, cefotaxime sodium 1 gm IV). The patient is appraised of the operative plan including the potential need for inotropic support and intraaortic balloon counterpulsation following VAD removal.

In the operating room a general anesthetic is administered. A Swan-Ganz catheter and radial arterial line are inserted. A transesophageal echocardiogram probe is inserted. Prior to proceeding with the operation hemodynamic indices and echocardiographic images should provide final confirmation that ventricular function has returned. The patient's chest, abdomen, and both anterior thighs including groins are prepped and draped into the operative field. Inotropes are mixed and available and an IABP and console are brought into the operating room for use as needed. VAD cannulae are prepped but draped out of the operative field in order to avoid mediastinal contamination.

The sternotomy is reopened and the mediastinal contents are swabbed for a gram stain, culture and sensitivity. VAD explantation is performed off CPB. Avoid manipulation of the heart as recovering ventricles are frequently irritable and cardiac compression can impede ventricular filling and create transient hypotension. Visual inspection of the free wall of the right ventricle and the anterior wall of the left ventricle provide the surgeon with some sense of the extent of ventricular recovery. Even though the heart is decompressed the presence of vigorous contractions suggests that ventricular recovery has occurred and the patient will tolerate removal of the mechanical blood pump. Heparin sodium (5000 U IV) is administered. If the patient is supported with an LVAD alone, LVAD flow is discontinued and the inlet and outlet cannulae clamped. The patient is observed for several minutes prior to decannulation to ensure that hemodynamic deterioration does not occur following discontinuation of LVAD support. Particular attention is paid to the trend in pulmonary artery pressures. A thermodilution cardiac output is obtained. Transesophageal echocardiography is used to evaluate left ventricular free wall and septal motion. A cardiotonic drug or two may be required, however, should more than one inotrope be required in moderate dose, it would be best to rest the ventricle for an additional day or two prior to LVAD removal. If the systemic blood pressure is maintained, cardiac output index exceeds 2.0 L/min/m², there is concentric contraction of the left ventricle by transesophageal echocardiography and pulmonary artery pressures do not rise following discontinuation of left ventricular assistance. Then left ventricular recovery is deemed complete (Table 5.8).

If left ventricular recovery has occurred the patient can be decannulated. Despite adequate anticoagulation during the period of mechanical circulatory support a rim of fibrin/thrombus occasionally develops around the inlet cannula at the point the inlet cannula enters the left atrium. If tension is placed on the pursestring sutures and the inlet cannula simply withdrawn any fibrin/thrombus at this location would be stripped from the inlet cannula with the potential for embolization. To reduce the chance for distal embolization, place positive pressure on the endotracheal tube and hold the pursestring sutures loosely as the inlet conduit is withdrawn. Such a maneuver allows a small amount of blood and any adjacent thrombotic material to be expelled at the time the inlet cannula is withdrawn. Concomitant positive airway pressure with the associated rise in left atrial

Table 5.8. Indicators of left ventricular recovery

In the absence of left ventricular assistance: Peak systolic arterial pressure	≥ 90 mm Hg
Pulmonary artery pressures do not rise.	
Cardiac output index	> 2.0 L/min/m²
Concentric contraction of left ventricular free wall and septum by transesophageal echocardiography.	

pressure also obviates the potential for air embolism. The outlet graft is clamped immediately adjacent to the aorta and the graft transected above the clamp. The outlet graft stump is oversewn with a running polypropylene suture. Alternatively, if the ascending aorta is generous and the patient tolerant of partial occlusion clamp application, the outlet graft may be excised in its entirety and the ascending aorta closed with a running row of polypropylene suture reinforced with short felt strips. If the patient has been supported with a centrifugal LVAD, the outlet cannula is withdrawn and the surrounding pursestring sutures tied. This maneuver is best performed with the patient in Trendelenburg position. Blood is permitted to eject from the aortotomy at the time the cannula is withdrawn. By so doing the potential for cerebral embolization of any fibrinous material on the outlet cannula is minimized.

5

If the patient is supported with two mechanical blood pumps, the RVAD is removed before the LVAD. After systemic heparinization right ventricular assistance is discontinued and the RVAD inlet and outlet cannulae clamped (Table 5.9). The patient is observed for several minutes to ensure that pulmonary artery pressures are maintained and right ventricular distention does not occur. In addition, LVAD filling is monitored as any degree of right ventricular failure is manifest by a decrease in LVAD flow. If right ventricular function is satisfactory the RVAD inflow cannula is withdrawn. Concerns about embolization are less critical than with extraction of the LVAD inflow cannula. The RVAD outlet cannula is transected adjacent to the anastomosis with the main pulmonary artery. The outlet cannula stump is oversewn. In the case of centrifugal pump right ventricular assistance, the outlet cannula is withdrawn from the main pulmonary artery and the surrounding pursestring sutures tied.

Attention is then directed to removal of the LVAD. LVAD removal is accomplished in a manner identical to that described above. As the time course for right ventricular recovery is shorter than the time course for left ventricular recovery,

Table 5.9. Indicators of right ventricular recovery

In the absence of right ventricular assistance: Peak systolic arterial pressure	$\geq$ 90 mm Hg
Pulmonary artery pressures are maintained.	
Central venous pressure	< 10-15 mm Hg
Cardiac output index	> 2.0 L/min/m^2
If an LVAD is in place, LVAD flow is maintained.	
Brisk contraction of right ventricular free wall.	
No more than one inotrope or pulmonary vasodilator required	

LVAD = left ventricular assist device

the situation occasionally arises in which the patient recovers right ventricular function before left ventricular recovery is complete. In this scenario, take the patient to the operating room and remove the RVAD leaving the LVAD in place, possibly decreasing the potential for sepsis. The amount of biomaterial within the mediastinum is reduced and the mediastinum irrigated. Following RVAD removal the sternum is closed and the patient allowed to recover for a few additional days until left ventricular function improves. At that time, the patient is returned to the operating room for LVAD explantation. Regardless of timing LVAD explantation is accomplished in a manner identical to that described above.

Once the patient has been decannulated, the cannulae are withdrawn from beneath the operative drapes. The mediastinum is copiously irrigated with warm, pulsed irrigation solution (neomycin sulfate-polymyxin B sulfate solution for irrigation, 1 ml in 1000 ml normal saline). All old thrombus and fibrin are carefully debrided and removed from the mediastinum. The mediastinum is drained with mediastinal tubes and closed in a routine fashion. A sterile dressing is applied. After the sternotomy incision has been covered by a sterile dressing the cannulae exit sites in the subcostal region are exposed. This is most readily accomplished by drawing back on the sterile drapes which have been previously applied to the prepped cannulae exit sites. The entry sites are debrided by saucerizing the skin edges with a scalpel and electrocautery. The underlying abdominal wall fascia is closed with absorbable suture. The overlying subcutaneous tissue and skin are closed primarily. If the cannula exit site was colonized during the period of mechanical circulatory support, the subcutaneous tissues and skin are left open and packed with a wet-to-dry dressing.

PATIENT CARE FOLLOWING VENTRICULAR ASSIST DEVICE REMOVAL

Following device explantation attention is once again directed to avoidance of complications. The potential for infection is minimized by routine perioperative antibiotics. The antibiotics are discontinued when mediastinal drains are removed. The patient is rapidly extubated and invasive hemodynamic monitoring lines removed as soon as the patient's hemodynamic and pulmonary status permit. Lines are rarely required beyond 24 hours following VAD removal. The patient is mobilized within 24 hours of surgery and aggressive cardiac rehabilitation follows. Device explantation is usually a brief operation and it is important that patient rehabilitation not be inordinately delayed by this return to the operating room. Provided ventricular recovery is complete prior to VAD removal, the patient can usually be transferred out of the intensive care unit within 24-48 hours of device explantation.

If the percutaneous cannulation sites are packed open, wound care is the only unique feature of the patient's subsequent hospitalization. The packing is replaced with a saline wet-to-dry dressing every 8 hours. If the patient's nutritional state has been optimized and the patient is sepsis free the wounds can be closed secondarily when a clean, granulating base appears. If there is any concern about occult sepsis, however, or the patient's nutritional status is suboptimal, wounds should be allowed to granulate closed. Depending upon the size of the defect secondary closure usually takes 4-8 weeks.

REFERENCES

1. Pae WE Jr. Ventricular assist devices and total artificial hearts: A combined registry experience. Ann Thorac Surg 1993; 55:295-298.
2. Jett GK. Postcardiotomy support with ventricular assist devices: Selection of recipients. Semin in Thorac and Cardiov Surg 1994; 6:136-139.
3. Guyton RA, Schonberger JPAM, Everts PAM et al. Postcardiotomy shock: Clinical evaluation of the BVS 5000 Biventricular Support System. Ann Thorac Surg 1993; 56:346-356.
4. Korfer R, El-Banayosy A, Posival H et al. Mechanical circulatory support with the Thoratec assist device in patients with postcardiotomy cardiogenic shock. Ann Thorac Surg 1996; 61:314-316.
5. Kieler-Jensen N, Lundin S, Ricksten SE. Vasodilator therapy after heart transplantation: Effects of inhaled nitric oxide and intravenous prostacyclin, prostaglandin E$_1$, and sodium nitroprusside. J Heart Lung Transplant 1995; 14:436-443.

5

Ventricular Assistance as a Bridge to Cardiac Transplantation

Wayne E. Richenbacher

INTRODUCTION

Mechanical ventricular assistance may be offered to end stage heart failure patients who fulfill cardiac transplant selection and exclusion criteria. The usual indication for mechanical circulatory support in the approved cardiac transplant candidate is hemodynamic decompensation. There is no expectation for myocardial recovery; rather the purpose of ventricular assist device (VAD) use in this patient population is to maintain systemic perfusion and allow the patient to be rehabilitated in preparation for the subsequent cardiac transplant. The VAD remains in place until a suitable donor is identified. The VAD is removed at the time of recipient cardiectomy. As newer electric devices employ a smaller controller and power source, system portability has improved. Patients who are receiving VAD support as a bridge to cardiac transplantation may now be discharged from the hospital to await the donor heart at home.

The first use of a mechanical blood pump, in this case a pneumatic artificial heart, as a bridge to cardiac transplantation was reported by Cooley and associates in 1969. A 47-year-old man suffering from end stage ischemic cardiomyopathy received 64 hours of circulatory support with the artificial heart prior to cardiac transplantation. Interest in cardiac transplantation waned during the 1970s as most recipients died within a few months of the transplant operation. With the discovery of cyclosporine A in 1976 the severity and acuity of rejection episodes decreased and steroid-sparing immunosuppressive protocols were developed. Survival rates improved, and as a result, there was a resurgence of interest in cardiac transplantation during the 1980s. As the number of cardiac transplants performed increased, the need for a device that was capable of interim support of the circulation was recognized. The concept of a mechanical bridge to transplantation evolved as a variety of devices were used to support patients with failing hearts. Carefully controlled clinical trials have led to Food and Drug Administration (FDA) approval of four different pulsatile VADs for use as a bridge to cardiac transplantation during the past 7 years. Currently, two-thirds to three-fourths of patients who require a mechanical blood pump as a bridge to cardiac transplantation survive to receive a donor heart. One year survival in patients who receive mechanical circulatory support prior to transplantation is 80-90%, a result that equals or exceeds the survival rate in patients who undergo cardiac transplantation alone.[1,2]

PATIENT SELECTION AND PRIMARY MANAGEMENT

6

The primary indication for use of a VAD as a bridge to transplantation is in a patient who fulfills cardiac transplantation selection criteria but who fails hemodynamically prior to the availability of a donor heart (Table 6.1). Patients with advanced congestive heart failure present with increased shortness of breath, dyspnea on exertion, paroxysmal nocturnal dyspnea, complaints of lethargy and easy fatigability. Physical findings include jugulovenous distention and peripheral edema. Cardiomegaly is manifest by lateral displacement of the palpated point of maximal impulse and pulmonary congestion on chest auscultation. Decreased breath sounds at the lung bases may indicate the presence of pleural effusions. Patients with advanced right heart failure develop abdominal distention with ascites and a palpable fluid wave. The liver edge may be palpated below the right costal margin. End stage right heart failure is manifest by a pulsatile liver margin. Laboratory abnormalities document the severity of the congestive heart failure. Fluid retention associated with congestive heart failure is accompanied by hyponatremia with serum sodium levels falling below 125-130 mEq/L. The low output state associated with advanced congestive heart failure leads to end organ dysfunction. End organ hypoperfusion is manifest by an elevation in serum urea nitrogen (BUN) and creatinine as well as a rise in liver function tests. Hepatic congestion secondary to central venous hypertension also leads to derangements of the clotting cascade including a prolongation in the prothrombin time.

Patients with congestive heart failure who deteriorate clinically are admitted to an intensive care unit and receive invasive hemodynamic monitoring. A right

---

Table 6.1. Cardiac transplant selection and exclusion criteria

End stage heart failure not amenable to other medical or surgical therapy.
Age ≤ 65 years.
No irreversible organ system failure.
Pulmonary hypertension, if present, is reversible.
No active infection.
Medically compliant, motivated for recovery.
Adequate financial resources or third party coverage.

heart catheterization determines the extent to which the pulmonary capillary wedge and central venous pressures are elevated. Initial treatment consists of inotropic therapy and an aggressive diuretic regimen. Inotropes of choice in this patient population include dopamine hydrochloride (2.5-15 µg/kg/min), dobutamine hydrochloride (2.5-15 µg/kg/min) and milrinone lactate (0.375-0.75 µg/kg/min). The efficacy of medical management is evidenced by increased urine output, negative fluid balance, a fall in cardiac filling pressures, rise in cardiac output, and ultimately, normalization of laboratory abnormalities. Associated with the improvement in hemodynamic parameters, the patient feels better and the physical examination documents an objective improvement in the patient's physical findings. Contemporary management of congestive heart failure utilizes the services of a cardiologist specifically trained in the management of patients with this disease. An appropriately staffed heart failure service also employs heart failure coordinators who maintain close telephone contact with such patients between frequently scheduled clinic visits. Subtle findings associated with advanced congestive heart failure precede full-blown clinical deterioration. Close, compulsive follow-up on behalf of the heart failure service allows the patient with heart failure to be hospitalized and managed on an urgent rather than emergent basis.

Should the patient continue to deteriorate despite aggressive medical management, intraaortic balloon pump (IABP) placement and mechanical ventricular assistance follow. In our opinion, as waiting times for cardiac transplantation have extended beyond weeks and months, the role of intraaortic balloon counterpulsation in this patient population has decreased. Intraaortic balloon counterpulsation effectively improves hemodynamics; however, the beneficial effect of intraaortic balloon counterpulsation can only be achieved for a few days at a time. More importantly, the efficacy of intraaortic balloon counterpulsation is limited by the associated risk of infection from a groin-based catheter and marked reduction in patient mobility associated with femoral arterial insertion. Our approach has been to utilize intraaortic balloon counterpulsation on a selective basis. If a patient experiences an acute deterioration in clinical condition, the IABP is employed for a few hours, a day or two at most, to allow time for preparations for VAD insertion. More commonly, hemodynamic deterioration is more gradual allowing time for urgent/semi-elective mechanical blood pump insertion. In the latter situation, intraaortic balloon counterpulsation is avoided altogether.

Evolving indications for mechanical circulatory support as a bridge to cardiac transplantation include unstable angina, and possibly, cardiac cachexia. Unstable angina can occur in the patient with ischemic cardiomyopathy and unreconstructable coronary artery disease. These patients have lost considerable myocardial function related to multiple past myocardial infarctions. Their intractable angina often results in an ongoing need for intravenous nitroglycerin. The patients are virtually bed bound as any activity exacerbates their myocardial ischemia. Narcotic requirements in these individuals is not insignificant. Although these patients may not fulfill the traditional hemodynamic criteria for VAD insertion, use of a mechanical blood pump in this clinical situation effectively reduces myocardial oxygen consumption thereby eliminating exercise induced myocardial ischemia. The presence of the mechanical blood pump eliminates the need for ongoing intravenous nitroglycerin. Pain management is improved and the patient is afforded the opportunity to rehabilitate. The second evolving indication for mechanical ventricular assistance as a bridge to cardiac transplantation, cardiac cachexia, occurs in a patient who has relatively stable end stage congestive heart failure but who is unable to rehabilitate due to a low functional class. Deconditioning is a real concern in this patient population as their inability to rehabilitate results in a dramatic loss of skeletal muscle tone and substance. As mechanical blood pump technology improves and devices become more portable, it is possible to envision a time when a mechanical blood pump can be inserted in patients who are simply incapable of exercising to any degree. Again, this patient population would not fulfill traditional hemodynamic selection criteria. However, the presence of the mechanical blood pump would allow this patient population to assume a more active lifestyle, rehabilitate and be better prepared for a subsequent heart transplant.

PREOPERATIVE PREPARATION

The preoperative evaluation and preparation of the patient for mechanical circulatory support as a bridge to cardiac transplantation begins at the time the patient is listed for transplantation. At the initial contact carefully evaluate the patient's family support network and home environment to ascertain whether or not the patient's psychosocial situation would permit mechanical circulatory support either inside or outside the hospital. As part of the financial review determine if the patient's provider has a policy concerning VAD coverage. As waiting times for cardiac transplantation have increased and the potential for deterioration in clinical condition arises, we also have a brief discussion regarding end stage therapy with the patient and his family at the time of the patient's formal transplant evaluation. We utilize the family meeting to educate the patient and his family about cardiac transplantation and possible management scenarios if the patient's clinical condition were to deteriorate. Included in this discussion are a few brief comments about the role of mechanical circulatory support. Interestingly, this discussion is often generated by the patient as they inquire about therapeutic alternatives should their clinical condition worsen. The transplant community is close knit and patients have often been introduced to the concept of mechanical ventricular assistance by other patients who have received or who currently require a VAD. We also utilize the Transplant Patient Support Group to discuss the concept of mechanical ventricular assistance at periodic intervals (see Chapter 9: Nursing Care of the Patient Requiring Mechanical Circulatory Support).

When a patient fulfills cardiac transplant selection and exclusion criteria he is listed with the United Network for Organ Sharing (UNOS) and assigned a status depending upon the acuity of his medical condition (Table 6.2). Patients who deteriorate hemodynamically and require in hospital care are assigned to UNOS Status 1. On January 20, 1999, the Status 1 designation was broken down into two

Table 6.2. United Network for Organ Sharing (UNOS) classification for cardiac transplant recipients

Status 1A
The patient must be an inpatient at the listing transplant center hospital and have at least one of the following devices or therapies in place.
1. Mechanical circulatory support for acute hemodynamic decompensation that includes one of the following:
   a. Left and/or right ventricular assist device implanted for 30 days or less
   b. Total artificial heart
   c. Intraaortic balloon pump
   d. Extracorporeal membrane oxygenator
2. Mechanical circulatory support for more than 30 days with objective medical evidence of a device related complication.
3. Mechanical ventilation.
4. Continuous infusion of at least a single high dose intravenous inotrope, in addition to continuous hemodynamic monitoring of left ventricular filling pressures.
5. The patient does not meet any of the criteria listed above, but has a life expectancy without a heart transplant of less than seven days.
Status 1B
The patient must have at least one of the following devices or therapies in place.
1. Left and/or right ventricular assist device implanted for more than 30 days
2. Continuous infusion of intravenous inotropes
Status 2
A patient who does not meet the criteria for Status 1A or 1B.
Status 7
A patient considered temporarily unsuitable to receive a transplant.

subcategories, Status 1A and 1B. Status 1A patients are those who are admitted to the listing transplant center hospital and who require mechanical circulatory support with a VAD for 30 days or less, total artificial heart, extracorporeal membrane oxygenation or intraaortic balloon counterpulsation. A patient is also assigned to Status 1A if the patient requires mechanical circulatory support with a VAD for more than 30 days but develops a device-related complication. Also included in this category are patients who require mechanical ventilation, continuous high dose inotrope infusion associated with continuous monitoring of left sided filling pressures and patients whose survival is expected to be less than 7 days. A patient is placed in the Status 1B category if the patient has had a VAD implanted for greater than 30 days or the patient requires continuous inotrope infusion without associated hemodynamic monitoring. Stable outpatients are assigned to UNOS Status 2.

When a patient is elevated to UNOS Status 1 we revisit the psychosocial and financial portions of the patient's initial transplant evaluation. With respect to financial coverage, we begin a dialogue with the patient's insurance company to confirm their policy toward coverage of mechanical blood pumps. If the provider has a no coverage policy or has not developed a coverage policy for mechanical circulatory support we take this opportunity to educate the provider about the role of ventricular assistance in the management of the potential cardiac transplant recipient. We stress the fact that patients who rehabilitate following VAD insertion will be better prepared for the subsequent heart transplant. We also educate the provider about the possibility of home discharge and its positive economic impact in this patient population.

The formal patient evaluation and preparation for mechanical blood pump insertion begins at the time the patient is admitted to the intensive care unit (Table 6.3). The patient must be reevaluated to ensure that no new contraindication to transplant has developed as the patient's clinical condition has deteriorated. The primary concerns are infection, pulmonary hypertension and end organ dysfunction. A white blood cell count is determined to ensure that the patient has not developed a leukocytosis or a left shift of the differential. A urinalysis and urine culture effectively rule out a urinary tract infection in a patient who frequently requires an indwelling urinary drainage catheter. A chest roentgenogram is inspected for evidence of pulmonary infiltrates. Intravenous sites, including invasive hemodynamic monitoring lines, are rotated to new locations at least every 72 hours and more frequently should the patient develop fever or an elevation in white blood cell count. It may have been some time since a right heart catheterization was performed. When a patient is admitted to the intensive care unit a Swan-Ganz catheter is inserted to update the right heart catheterization pressures. If there has been an interval increase in pulmonary artery pressures, the patient receives a trial of inotropic therapy or intravenous nitroglycerin with associated diuresis to determine the reversibility of the pulmonary hypertension. The presence of pulmonary hypertension, particularly when fixed, will have an impact

---

Table 6.3. Preoperative checklist for ventricular assist device implantation

1. Baseline laboratory studies
   a. Complete blood count
   b. Renal panel
   c. Liver function tests
   d. Coagulation studies
   e. Serum prealbumin
   f. Type and cross (leukodepleted blood products)
2. Right heart catheterization
   a. Determine reversibility of pulmonary hypertension
3. Informed consent
4. Echocardiography
   a. Left ventricular thrombus
   b. Aortic insufficiency
   c. Patent foramen ovale
5. Baseline neurologic examination
6. If the patient has had a previous open heart operation review
   a. Old operative note(s)
   b. Most recent cardiac catheterization
7. New intravenous and hemodynamic monitoring lines the night prior to surgery
8. Preoperative antibiotics and skin prep
9. Ensure mechanical blood pump team availability
   a. Operating room
   b. Surgical intensive care unit

upon the conduct of the operation to insert the VAD. The presence of fixed pulmonary hypertension may also influence device selection as it may be predictive of the need for biventricular support. End organ function is evaluated by determining serum BUN and creatinine levels as well as a liver function panel. If the liver function tests are increased there may be problems with synthetic function. In this situation it is prudent to check a coagulation panel as the prothrombin time is frequently elevated.

Proceed with preoperative planning and preparation provided the patient still meets cardiac transplant selection criteria, has deteriorated to the point where he will not survive without mechanical circulatory support and he has not developed a contraindication to mechanical circulatory support (Table 6.4). A frank discussion is held with the patient and family regarding the risks and benefits of mechanical circulatory support (Table 6.5). It is important to include the patient's family in this preoperative discussion as the patient may have an altered sensorium due to the low cardiac output state. At the time informed consent is obtained we describe the benefits of mechanical circulatory support as the ability to stabilize the patient's hemodynamics and improve cardiac output. The patient

Table 6.4. Selection criteria for mechanical circulatory support as a bridge to cardiac transplantation

1. Patient fulfills cardiac transplant selection criteria.
2. Hemodynamic deterioration necessitating use of inotropes and, possibly, intraaortic balloon counterpulsation.
3. Patient fulfills "traditional" hemodynamic criteria.
   - a. Cardiac output index < 2.2 L/min/m^2
   - b. Systemic arterial blood pressure ≤ 100 mm Hg
   - c. Pulmonary capillary wedge pressure $\geq 15\text{-}20$ mm Hg

Table 6.5. Informed consent

Risks and benefits of mechanical circulatory support as a bridge to cardiac transplantation

Risks
1. Bleeding
2. Infection
3. Stroke
4. Irreversible end organ dysfunction
5. Device malfunction
6. Device dependency (contraindication to transplant)
7. Death

Benefits
1. Improved systemic perfusion
2. Reversal of end organ dysfunction
3. Permit rehabilitation
4. Improved quality of life (possibility of home discharge)
5. Allow survival to cardiac transplantation

and family are informed that this increase in cardiac output usually results in improved end organ function and permits the patient to rehabilitate prior to cardiac transplantation. The presence of the mechanical blood pump may also improve the patient's potential for survival to transplant and will have a positive impact upon the patient's quality of life as he will be allowed out of the intensive care unit and may be discharged to home. Risks associated with the use of the mechanical blood pump as described to the patient and family include bleeding and the possible need for a blood transfusion, infection related to the sternotomy and presence of biomaterial, in particular, the percutaneous cannulae or drive line. We impress upon the patient and family members that the infection risk is ongoing due to the presence of the percutaneous cannulae or drive line. The possibility of stroke and irreversible end organ dysfunction are noted. The patient and family are informed that there is the possibility for device malfunction. Most importantly, the patient and family are made aware that should the patient develop a contraindication to transplant during the period of mechanical blood pump support he may be removed from the cardiac transplant waiting list. Should the patient develop a complication during the period of mechanical circulatory support or become device dependent with a contraindication to transplantation the patient is informed that death is a possibility.

After informed consent is obtained, the patient is typed and crossed for 4 units of packed red blood cells. Inform the blood bank that the patient is a potential transplant recipient as the blood products should be leukocyte depleted to avoid antibody development. Baseline laboratory studies in preparation for surgery include a complete blood count, liver function tests, renal panel and coagulation studies. A serum prealbumin provides a reasonable estimate of the patient's nutritional state. If the patient is receiving continuous heparin sodium therapy, a platelet count is determined to ensure that heparin associated thrombocytopenia has not occurred. An echocardiogram is performed. This can either be a transthoracic echocardiogram or transesophageal echocardiogram depending upon the adequacy of windows. Specific information to be derived from the echocardiogram includes an evaluation of the left ventricular apex, the status of the aortic valve and identification of a patent foramen ovale. The left ventricular apex is a frequent site for thrombus deposition in a patient with a low ejection fraction. If a left ventricular thrombus is present, it is important to avoid manipulating the left ventricle prior to placement of the aortic crossclamp to avoid a thromboembolic event at the time the VAD is inserted. It is also important to remove all thrombus from the left ventricle at the time of apical cannulation to avoid subsequent thromboembolic events. The aortic valve is inspected for competency. The presence of aortic insufficiency will compromise systemic perfusion following VAD placement as a percentage of VAD flow will pass retrograde across a regurgitant aortic valve into the patient's left ventricle where it will reenter the VAD. If a patient has aortic insufficiency, aortic valve replacement or obliteration of the left ventricular outflow tract should be considered at the time of VAD insertion (see Chapter 7: Special Situations). A bubble study, as described in Chapter 5, is performed to identify a patent foramen ovale. If the foramen ovale is patent, a right-to-left shunt with systemic desaturation may occur following left VAD (LVAD) insertion. If a patent foramen ovale is identified the patient should receive bicaval cannulation rather than a single venous cannula for cardiopulmonary bypass (CPB) at the time of blood pump insertion. This cannulation scheme allows a right atriotomy and foramen ovale closure as described below. Unfortunately, due to marked elevation in atrial pressures on both sides of the atrial septum the patent foramen ovale is difficult to identify preoperatively. Thus, transesophageal echocardiography should be repeated in the operating room after left ventricular assistance has been initiated.

A baseline neurologic examination is performed. Due to the potential for thromboembolic events following VAD insertion, it is helpful to identify subtle neurologic findings prior to VAD insertion. An abdominal examination is repeated. Any reported abdominal pain, distension or tenderness on abdominal palpation should prompt a work-up for ischemic gut, an occasional sequela of a low output state. The abdomen is also inspected for scars from previous abdominal surgery. Carefully choose drive line and cannulae exit site locations avoiding such scars if present. If a patient has had a previous sternotomy and cardiac operation, a lateral chest roentgenogram is reviewed to see if there is a clear space beneath the sternum (Fig. 6.1). As the patient with heart failure frequently has a distended right ventricle it is possible to injure the heart at the time of repeat sternotomy. Having said that, however, we attempt to place cannulae in the chest and avoid groin cannulation for CPB as the common femoral artery and vein will need to be cannulated at the time of VAD explantation and cardiac transplantation. If a patient has

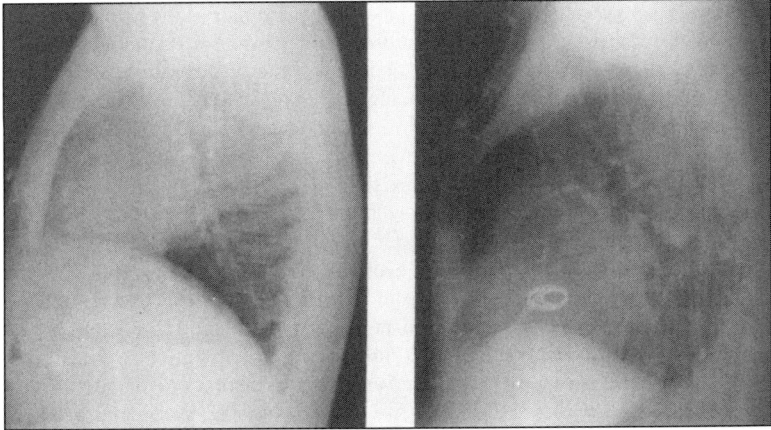

Fig. 6.1. Operative planning for blood pump insertion in the patient who has had a previous open heart operation includes a review of the lateral chest roentgenogram. A. Mediastinal contents are opposed to the posterior table of the sternum. In this instance consider exposing the femoral vessels prior to sternotomy. B. The clear space beneath the sternum suggests that a sternotomy can be performed with little risk of injuring the heart or great vessels.

had previous open heart surgery, take time to review the old operative note to identify the location of bypass grafts, note the previous surgeon's comments about aortic length, the presence or absence of aortic mural calcification and whether or not the pericardium was left open or closed. If a patient has had previous coronary revascularization, review the most recent cardiac catheterization to determine graft location and patency, in particular, internal mammary artery grafts which may approach the midline and be injured at the time of repeat sternotomy.

Perioperative antibiotics include vancomycin hydrochloride (15 mg/kg IV 1 hour preoperatively and every 12 hours postoperatively for 48 hours), levofloxacin (500 mg IV 1 hour preoperatively and every 24 hours postoperatively for 48 hours), rifampin (600 mg PO 1-2 hours preoperatively and daily postoperatively for 48 hours) and fluconazole (200 mg IV preoperatively and every 24 hours postoperatively for 48 hours). Skin preparation includes a chlorhexidine gluconate (Hibiclens; Zeneca Pharmaceuticals, Wilmington, DE) shower the night prior to surgery. If the acuity of the patient's condition or presence of invasive hemodynamic monitoring lines do not permit a shower, a bedside scrub should be performed. Prior to taking the patient to the operating room the body hair should be clipped from the chin to the knees, bedside to bedside. Ensure that the mechanical blood pump team is available in the operating room and surgical intensive care unit. Most importantly, VAD insertion as a bridge to transplantation should be performed on an urgent not emergent basis. In our opinion, there is no reason to implant a VAD as a bridge to transplant on an emergent basis as this simply means that the patient's condition was allowed to deteriorate too far prior to considering this therapeutic modality. The rate of deterioration in heart failure patients is usually gradual enough that semi-elective VAD insertion can be planned allowing adequate time for preparation of both the patient and family as well as the mechanical blood pump team.

VENTRICULAR ASSIST DEVICE (VAD) SELECTION

Four devices are currently approved by the FDA for use as a mechanical bridge to cardiac transplantation (Table 6.6). The Thoratec VAD (Thoratec Laboratories Corp., Pleasanton, CA) is a paracorporeal, pneumatically driven pulsatile device that is also approved for use in postcardiotomy cardiogenic shock. This device is available with a variety of inlet and outlet cannulae that permit the Thoratec blood pump to be configured for either right or left heart support. Two Thoratec VADs can be utilized for biventricular assistance. The extracorporeal nature of the Thoratec blood pump makes this the ideal VAD for the diminutive patient with a small body habitus. We employ the Thoratec VAD in the patient who has a body surface area < 1.5 m². Recently, Thoratec described the use of their device to provide left heart support in a 17 kg, 7-year-old patient with a body surface area of 0.73 m².[3] This approach eliminates technical problems associated with insertion of a large implantable mechanical blood pump in a confined space. Even if an implantable VAD can be inserted into a small patient, the device can impact upon

Table 6.6. Ventricular assist device (VAD) selection

Thoratec VAD System
     Patient body surface area < 1.5 m^2
     Previous upper abdominal surgery
     Potential for right heart failure requiring mechanical right ventricular assistance
Thermo Cardiosystems HeartMate 1000 IP LVAS
     Patient's home is a great distance from the hospital
     Patient's social situation precludes hospital discharge
     Anticipated duration of VAD support is brief (small patient, blood type AB)
Thermo Cardiosystems HeartMate VE LVAS
Novacor N-100 LVAD
     Patient lives within a few hours of the hospital (by ground or air)
     Patient's social situation permits hospital discharge
     Anticipated duration of VAD support is long (large patient, blood type O)

6

the patient's costal margin creating chronic pain and decreased mobility or compress abdominal viscera creating early satiety. We also utilize the Thoratec VAD if previous upper abdominal surgery or a surgical incision in the left subcostal region precludes preperitoneal or intraperitoneal blood pump placement. Preoperatively we try to develop some sense of the potential for post-VAD insertion right heart failure. If the patient has markedly elevated pulmonary artery pressures, usually exceeding 70-80 mm Hg, with minimal reversibility following the administration of sodium nitroprusside or nitroglycerin, the likelihood that the patient will require both left and right heart support is increased. As the Thoratec VAD can also be configured as a right VAD (RVAD) we would employ the Thoratec VAD in this clinical scenario. Avoidance of hybrid biventricular support (an implantable LVAD and Thoratec RVAD) greatly simplifies the patient's postoperative management.

Until recently, the only intracorporeal device approved by the FDA for use as a bridge to cardiac transplantation was the Thermo Cardiosystems HeartMate 1000 IP LVAS (Thermo Cardiosystems, Inc., Woburn, MA) pneumatically powered LVAD.[4] The Thermo Cardiosystems pneumatic LVAD is designed for left ventricular apex cannulation and can only be configured for left heart support. The device is intracorporeal and is positioned in either the left upper quadrant of the patient's abdomen or the preperitoneal position in the left subcostal region. The device is connected by a percutaneous drive line to an external drive unit. The drive unit is smaller and based on large wheels which makes this drive unit more portable than the heavier Thoratec pneumatic drive unit. Patients supported with both the Thermo Cardiosystems 1000 IP LVAS and the Thoratec ventricular assist system are currently required to remain hospitalized. In general, we only use the pneumatic version of the Thermo Cardiosystems VAD when the patient's home is a great distance from the hospital or the patient's social situation precludes hospital discharge. The latter scenario might include a patient who lives alone, who has no family members or support network that can assist with drive line care or a patient who is simply uncomfortable with managing a VAD and associated hardware in an out of hospital setting.

In 1998, the FDA approved two electrically powered LVADs for clinical use as a bridge to cardiac transplantation. The Thermo Cardiosystems HeartMate VE LVAS (Thermo Cardiosystems, Inc., Woburn, MA) and the Novacor N-100 LVAD (Novacor Division, Baxter Healthcare Corp., Santa Ana, CA) are both implantable blood pumps designed solely for left heart support. Both VADs employ a motor that is contained within the mechanical blood pump. The electric motor is powered by an external controller and battery pack. The implanted blood pump is connected to the external power source by a percutaneous hard wire. The obvious advantage of the electric implantable LVAD is dramatically improved portability. More importantly, patients supported with the Thermo Cardiosystems HeartMate VE LVAS and the Novacor N-100 LVAD are permitted to be discharged from the hospital. The disadvantage of using a blood pump that is only designed for left heart support is the occasional patient that requires concomitant right ventricular assistance.

In an effort to maximize patient mobility and provide the patient with the option for home discharge, we insert an electric VAD whenever possible. Our University Hospital is in a rural setting; thus, patients can live a considerable distance from not only our hospital but also any healthcare facility. When making the VAD selection we take into consideration the location of the patient's home with respect to the hospital as well as the patient's social situation at home. We prefer that patients live within 2-3 hours travel time to the hospital. This allows the patient to quickly return to the hospital in the event of a pump or other medical emergency. By living in close proximity to the hospital the patient also has ready access to the outpatient clinic for frequent follow-up visits. In reality, patients may live much greater distances from the hospital, as a pump or other medical emergency justifies the use of air transportation which greatly reduces travel times. The need for close follow-up through the outpatient clinic probably represents a greater burden to the patient who lives a distance from the hospital. This concern will undoubtedly lessen over time as referring cardiologists become more comfortable with patients requiring outpatient mechanical circulatory support. Video conferencing techniques through the patient's home or local healthcare facility will also enable the tertiary care center to maintain close contact with both the patient and referring physician without the need for frequent, routine clinic visits.

OPERATIVE MANAGEMENT

PREIMPLANTATION PREPARATION
In addition to the preoperative skin preparation and antibiotic coverage, all invasive hemodynamic monitoring lines are removed and replaced the night prior to surgery to minimize infectious complications. Full dose aprotinin is administered if CPB is to be employed for VAD insertion (Table 5.3). Aprotinin is also

employed at the time of VAD explantation and cardiac transplantation. The potential for an aprotinin reaction with the second administration exists; however, the risk of bleeding at VAD implantation exceeds the risk of an aprotinin reaction at the second operation. Management considerations at the second operation are described below.

The pulsatile devices employed as a bridge to cardiac transplantation require a significant amount of preparation time (see Chapter 11: Device Specific Considerations). Operating room nurse coverage is increased and adequate time allowed for device preparation prior to implantation. At our institution a routine open heart operation utilizes the services of one scrub nurse and one circulating nurse. When a VAD is implanted one scrub nurse is responsible for blood pump preparation at a separate back table. A second operating room scrub nurse is positioned at the operating room table while a third nurse serves as a circulator.

If the patient has had a previous sternotomy the usual precautions are taken at the time the sternum is reopened. Blood that is appropriately typed, crossed and leukodepleted is immediately available in the operating room. Heparin sodium is drawn up and the anesthesiologist is prepared to administer the heparin sodium should emergent CPB be required. Femoral arterial and venous cannulae are available although the groin is only opened if the potential for technical misadventure is high. The femoral vessels are exposed in patients in whom the right ventricle is juxtaposed to the posterior table of the sternum on lateral chest roentgenogram or in whom a patent internal mammary artery graft crosses the midline. Ideally, the groin is left inviolate as femoral arterial and venous cannulation are employed at the time of mechanical blood pump explantation and cardiac transplantation. The CPB machine is primed and available in the operating room with perfusion personnel in attendance.

A standard sternotomy is performed and the pericardium suspended. A left atrial pressure line is inserted via the right superior pulmonary vein. It is easier to insert this line immediately after CPB is initiated and the heart decompressed. If CPB is not employed the left atrial line is inserted prior to VAD cannulation. As the sternum is opened and preparations for cannulation are in progress transesophageal echocardiography is repeated. The purpose of transesophageal echocardiography at this point is to carefully inspect the atrial septum to identify a patent foramen ovale. If the atrial septum is clearly intact a single two-stage CPB venous cannula is employed. The cannula is inserted through the right atrial appendage. If there is any suggestion that the foramen ovale is patent bicaval cannulation is employed.

THORATEC LVAD INSERTION

Device specific preparation and implantation techniques are outlined in Chapter 11: Device Specific Considerations. Ventricular apex cannulation is the preferred inflow route for left ventricular assistance as the left ventricle is more effectively decompressed maximizing LVAD flow and systemic perfusion when compared to left atrial cannulation. Left atrial cannulation, although rarely employed in the patient being bridged to transplantation, may be the inflow site of choice in the patient who has suffered a recent anterior wall myocardial infarction complicated by cardiogenic shock. Patients who have suffered an acute anterior wall myocardial infarction frequently develop thrombus within the left ventricular apex. More importantly, the acutely infarcted left ventricular apex may be necrotic. Placement of an inflow cannula into a necrotic left ventricular apex is tenuous at best as bleeding from the cannulation site is common and there is always the potential for apical disruption following cannula placement. If the left ventricular apex is deemed unsuitable for cannulation, it may be possible to insert the Thoratec left atrial cannula without placing the patient on CPB. Whether or not CPB is required is determined by the extent of right atrial dilation. If the right atrial free wall can be retracted to expose the right superior and inferior pulmonary veins it may be possible to insert the left atrial cannula without the need for CPB. If, however, exposure of the junction of the right sided pulmonary veins and left atrium creates hemodynamic instability, it is safer to place the patient on CPB and decompress the heart while this maneuver is accomplished.

The points at which the inlet and outlet cannulae will traverse the subcostal region are identified and the skin buttons removed and fascial defects created prior to heparinization. For left atrial cannulation the inlet and outlet cannulae transverse the skin in the right subcostal region (Fig. 6.2). The LVAD inflow cannula is positioned laterally. The LVAD outflow cannula is positioned medially. If left atrial cannulation is to be employed the left atrial pressure line is inserted into the right inferior pulmonary vein in order to preserve the junction between the right superior and inferior pulmonary veins for atrial cannulation. Waterston's groove is developed and the pursestring sutures and atrial cannulation site located on the wall of the left atrium thereby avoiding an obstruction to flow through the right superior pulmonary vein (Fig. 5.7). If the patient is to be cannulated without the

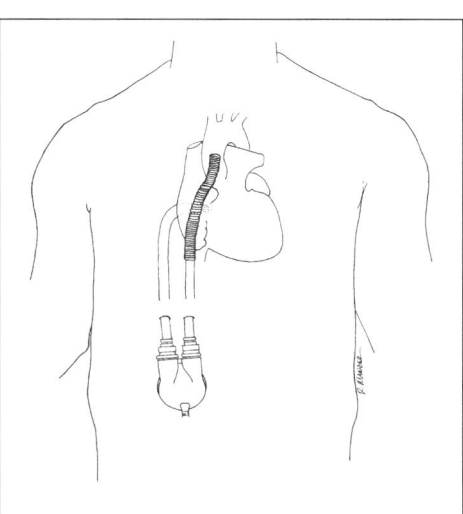

Fig. 6.2. Cannulation scheme for Thoratec left ventricular assistance using a left atrial inflow cannula.

use of CPB heparin sodium (5000 U IV) is administered. The inlet cannula is passed through the subcutaneous tunnel, inserted into the left atrium and the pursestring sutures snared and tied over buttons. Outlet cannula placement follows. The outlet graft is anastomosed to the right anterolateral aspect of the ascending aorta. The use of a partial occlusion clamp allows placement of the outlet graft in this location. In the cardiomyopathic patient, the partial occlusion clamp should be applied to the aorta and the patient's hemodynamic status observed for several minutes prior to creating the aortotomy. The increase in left ventricular afterload created by the partial occlusion clamp may result in left ventricular dilation and decompensation. If the patient is intolerant of partial occlusion clamp application, CPB should be employed for the LVAD outflow graft-to-aortic anastomosis. In the postcardiotomy cardiogenic shock patient, the location of the outlet graft-to-aortic anastomosis is dictated by the presence of proximal vein graft anastomoses and cannulation sites. In the bridge to transplant patient, the outlet graft anastomosis should be located as low on the aorta as possible. By placing the outlet graft in close proximity to the root of the aorta the outlet graft remnant can be removed at the time of recipient cardiectomy and cardiac transplantation. The goal is to minimize the amount of biomaterial remaining behind in a patient who undergoes cardiac transplantation. After the outlet graft has been sewn onto the ascending aorta, the vascular clamp is moved from the aorta to the graft so the anastomosis may be checked for hemostasis. The outlet graft is passed through the subcutaneous tunnel.

Left ventricular apex cannulation requires CPB to decompress the left heart and allow the surgeon to elevate the left ventricle into the operative field. Whether or not an aortic crossclamp is applied and cardioplegia administered is a matter of preference. We prefer a brief period of cardioplegic arrest as it affords a better opportunity to place the apical cannula at the tip of the left ventricular apex and allows an unobstructed view of the left ventricular endocardial surface. Careful inspection of the inside of the left ventricle ensures that thrombus, if present, is completely removed from the left ventricular apex and the inflow cannula is adequately positioned within the body of the left ventricular cavity. The author does not believe that a brief period of cardioplegic arrest leads to a decline in right ventricular performance.

Once again, skin buttons, fascial defects and percutaneous tunnels are created prior to heparinization. For left ventricular apex cannulation the cannulae traverse the skin in the left subcostal region. The inflow cannula is in the lateral position while the outflow cannula is located medially (Fig. 6.3). CPB is initiated and normothermia maintained. The outlet graft is anastomosed to the ascending aorta first. Left ventricular apex cannulation follows. After the outlet graft has been anastomosed to the ascending aorta and the suture line checked for hemostasis, the aorta is crossclamped and cardioplegia delivered into the aortic root. The heart is elevated and several iced packs placed behind the heart to facilitate exposure of the left ventricular apex. The two most common errors in placing a left ventricular apex cannula include coring the left ventricular apex too close to the septum or placing the diaphragmatic tunnel too far toward the left lateral aspect of the pericardium. With the heart decompressed it helps to invaginate the left ventricular apex to identify the ventricular septum prior to removing the core at the left ventricular apex. Avoid placing the left ventricular apex cannula into the septum with resultant inflow cannula obstruction. If the patient's heart is markedly dilated the left lateral aspect of the pericardium may move far into the left chest, often approaching the left lateral chest wall. If the diaphragmatic tunnel is created at the left ventricular apex, inflow obstruction will occur when left ventricular assistance is initiated and the heart is decompressed (Fig. 6.4). This problem can be avoided by creating a diaphragmatic tunnel approximately two-thirds of the way between the midline and the left lateral extent of the pericardium. When left ventricular assistance is initiated the heart tends to move medially as the left ventricle is decompressed. This allows the left ventricular apex cannula to lie within the

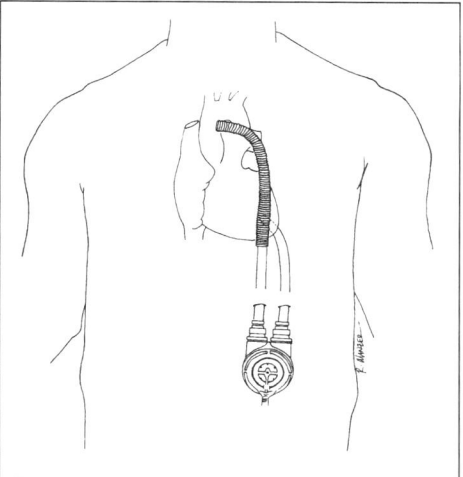

Fig. 6.3. Cannulation scheme for Thoratec left ventricular assistance using a left ventricular apex inflow cannula.

6

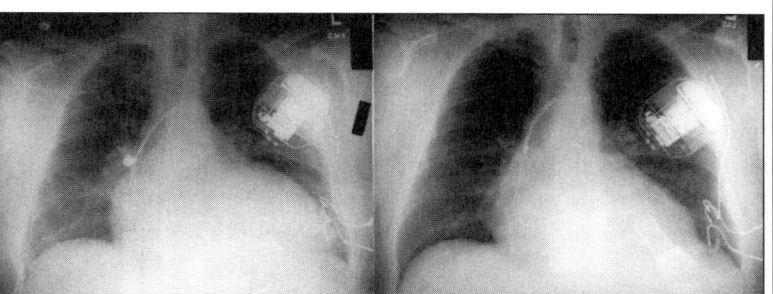

Fig. 6.4. A. Massive cardiomegaly seen on a preoperative chest roentgenogram. B. Left ventricular-to-aortic left ventricular assistance effectively decompresses the heart and reduces the transverse dimension of the cardiac silhouette.

body of the left ventricular cavity. The core is removed from the left ventricular apex, the apical cannula inserted and passed through the diaphragmatic and percutaneous tunnels. The LVAD is deaired as it is connected to the inlet and outlet cannulae.

Thermo Cardiosystems or Novacor LVAD Insertion

If an implantable LVAD is employed the sternotomy is followed by creation of a preperitoneal pocket and drive line tunnel. The patient is not anticoagulated until the surgical dissection is complete to minimize the potential for postoperative bleeding. We prefer a preperitoneal pocket to intraperitoneal implantation as there is no postoperative ileus and the possibility of an intraabdominal complication is virtually eliminated. Development of the preperitoneal pocket begins in the midline (Fig. 6.5). The left rectus abdominis muscle is elevated from the posterior rectus sheath. As the dissection continues laterally the abdominis oblique and transversus abdominis muscles are elevated from the transversalis fascia. The fascia is taken down from the caudad surface of the diaphragm and a diaphragmatic tunnel created as described above. Care is taken to achieve hemostasis within the pocket. Defects in the peritoneum are identified and repaired. The drive line must pass from the pocket through the abdominal wall muscles, soft tissue and skin. A skin button is created in the appropriate location and the drive line tunnel formed. The pocket and tunnel are packed with a vancomycin hydrochloride (1 gm in 1000 ml normal saline) soaked sponge. At this point the patient is heparinized.

If aortic length and diameter permit, anastomose the outlet graft to the ascending aorta prior to initiating CPB. This will minimize CPB time. Otherwise

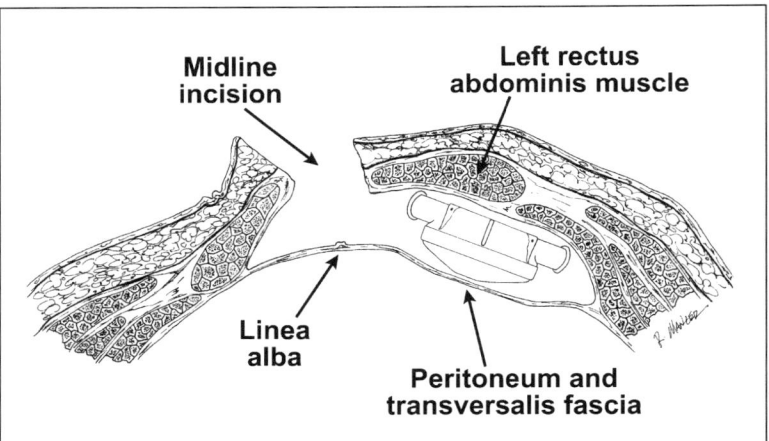

Fig. 6.5. The preperitoneal pocket is created in the left upper quadrant of the patient's abdomen. The plane of dissection is located between the abdominal wall musculature and the transversalis fascia. The right rectus abdominis is elevated to permit unobstructed passage of the outlet graft.

institute CPB and perform the outlet graft-to-aortic anastomosis with either a partial occlusion clamp or aortic crossclamp and cardioplegic arrest. Left ventricular apex cannulation follows (Fig. 6.6). The VAD is inserted into the preperitoneal pocket as the drive line is passed through the abdominal wall tunnel. The inlet cannula is inserted into the heart and the left heart, LVAD and cannulae are deaired as the outlet graft is attached to the blood pump. Deairing is facilitated by placing the patient in steep Trendelenburg and reducing CPB flow thereby allowing the heart to fill. The left heart is gently shaken as the left atrial appendage is inverted and positive pressure placed on the endotracheal tube. Blood will emanate from the mechanical blood pump. Transesophageal echocardiography aids in the identification of air bubbles within the left atrium and ventricle. After the outlet graft has been connected to the LVAD, the outlet graft and aortic root are vented until the left heart is completely deaired. The mechanical blood pump is connected to the appropriate drive mechanism and the pump pulsed in a low, fixed rate mode. Continuous aspiration on the aortic root vent ensures that any final air bubbles ejected from the blood pump itself will be removed from the systemic circulation.

TERMINATION OF THE VAD IMPLANT OPERATION

Once deairing is deemed complete the patient is returned to the supine position and all vents removed. CPB flow is serially decreased as LVAD flow is increased. Both the central venous and left atrial pressures are simultaneously monitored as they allow the surgeon to determine the source of the problem should LVAD flow be inadequate (Table 5.5). The goal is a cardiac (LVAD flow) index in excess of 2 L/min/m^2 with adequate LVAD filling and a low left atrial pressure (Fig. 6.7). Transesophageal echocardiography is repeated following the discontinuation of CPB (Table 6.7). A bubble study is performed as the atrial septum is

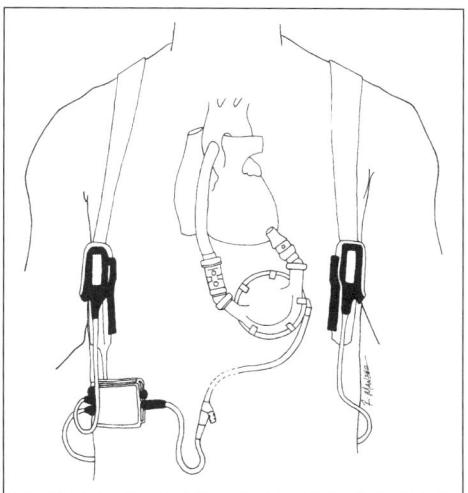

Fig. 6.6. Cannulation scheme for Thermo Cardiosystems VE LVAS left ventricular assistance.

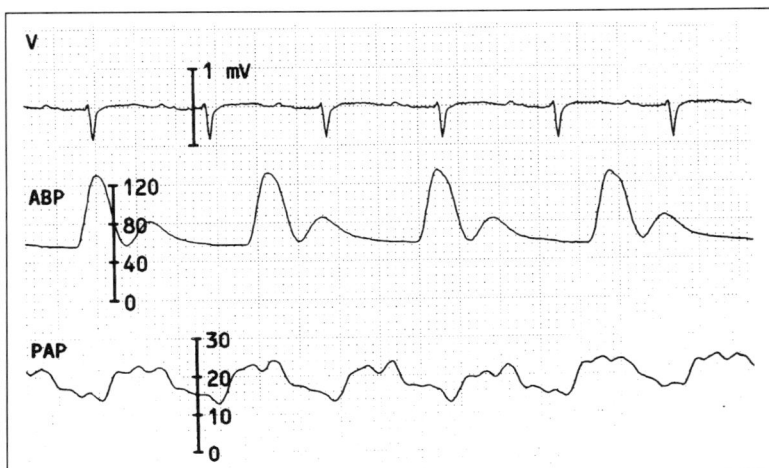

Fig. 6.7. Simultaneous electrocardiogram (V), systemic arterial (ABP) and pulmonary arterial (PAP) pressure traces in a patient receiving left ventricular assistance. There is a 1:1 correlation between the electrocardiogram and pulmonary arterial pressure trace as native right heart ejection perfuses the lungs. There is asynchrony between the electrocardiogram and systemic arterial pressure trace. There is no native left ventricular ejection. Systemic perfusion is entirely dependent upon left ventricular assist device ejection.

Table 6.7. Transesophageal echocardiography in the patient receiving left ventricular assistance

---

1. Atrial septum
    Presence/absence of patent foramen ovale.
2. Left atrium
    Is the left atrium well decompressed?
    Is the left atrial cannula well positioned?
3. Left ventricle
    Is the left ventricle well decompressed?
    Is the left ventricular apex cannula well positioned?
    Is there free flow of blood into the apical cannula?
    Does the aortic valve open?

---

inspected one final time. As the left atrial pressure has been markedly reduced, flow across a patent foramen ovale is easily recognized. If a patent foramen ovale is identified, regardless of whether it is associated with systemic desaturation, CPB is reinstituted using bicaval cannulation. LVAD pumping is briefly discontinued. Caval snares are applied and the patent foramen ovale closed through a small right atriotomy. The left atrium or ventricle is inspected to ensure adequate placement of the left atrial or left ventricular apex cannula, respectively. The relative location of the left ventricular apex cannula with respect to the left ventricular free wall and ventricular septum is documented. If the left ventricular apex cannula is adequately positioned there should be a free flow of blood from the left ventricle into the left ventricular apex cannula. The left ventricle should be well decompressed and the native aortic valve should not open at any point in the cardiac cycle.

Right ventricular function is assessed.[5] If the central venous pressure exceeds 15 mm Hg in association with a low left atrial pressure and an inadequate cardiac output, index specific therapies for right ventricular failure are initiated. Isoproterenol hydrochloride (0.01-0.20 μm/kg/min IV) is administered, rhythm permitting. Nitric oxide (20-60 ppm) usually has a profound positive impact on right ventricular dysfunction. If a patient has right ventricular failure refractory to inotropic therapy or nitric oxide administration an RVAD should be inserted. The only device approved for use as an RVAD as a bridge to cardiac transplantation is the Thoratec VAD. The inlet cannula is inserted through concentric pursestring sutures in the right atrial free wall while the outlet graft is anastomosed to the main pulmonary artery. The latter anastomosis is accomplished as close to the pulmonic valve as possible. This will allow complete removal of the main pulmonary artery and outlet graft at the time of cardiac transplantation. The cannulae are exteriorized to the side opposite the LVAD cannulae. If the patient is supported with a Thoratec LVAD using a left atrial inflow cannula the RVAD cannulae are brought through the skin in the left subcostal region. The RVAD inflow cannula is medial. The RVAD outflow cannula is lateral. The RVAD inflow cannula must cross the LVAD outflow cannula anterior to the heart. If the patient is supported with an LVAD using a left ventricular apex cannula, regardless of manufacturer, the RVAD cannulae exit the chest in the right subcostal region (Fig. 6.8).

6

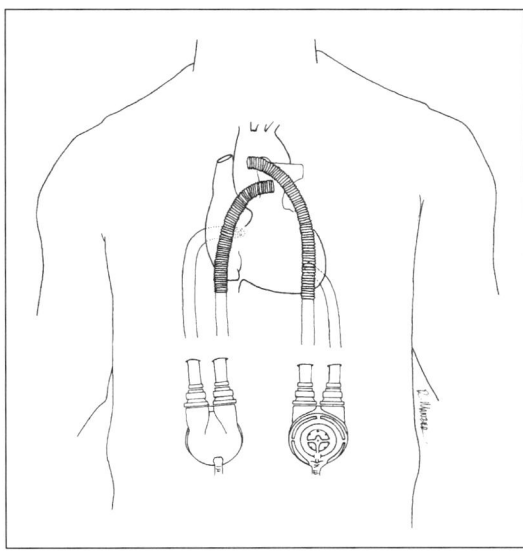

Fig. 6.8. Cannulation scheme for Thoratec biventricular assistance. The left ventricular assist device employs a left ventricular apex cannula.

In this instance the RVAD outflow cannula will cross the LVAD outflow cannula anterior to the heart. It is usually necessary to place the patient back on CPB during RVAD insertion. Right heart decompression eliminates the need for pursestring suture placement in a distended right atrium. Pursestring suture placement in a thin walled right atrium with a central venous pressure exceeding 20 mm Hg can result in severe postoperative bleeding.

Other than drug therapy for right heart support, the patient should not require inotropic therapy for left ventricular support. If LVAD flow is inadequate it is important to evaluate the position of the LVAD inflow cannula, right ventricular contractility and the patient's volume status prior to removing the CPB cannulae and closing the patient's chest. Hypovolemia is easily corrected by transfusion from the CPB machine. Right ventricular failure is treated as described above. If LVAD flow is inadequate due to a malpositioned inflow cannula, this problem should be rectified by reorienting the cannula using transesophageal echocardiographic guidance.

Protamine sulfate is administered and the effect of heparin sodium completely reversed. The activated clotting time (ACT) should return to baseline level. Blood product administration is directed by specific laboratory abnormalities of clotting parameters. Minimize blood product administration to reduce the antigen load in a patient who will ultimately require cardiac transplantation. The pericardium and left upper quadrant preperitoneal pocket, if present, are irrigated with vancomycin hydrochloride solution (1 gm in 1000 ml normal saline). The outlet graft from an implantable LVAD is located off the midline to lessen the chance of injuring this conduit at the time of sternal reentry and cardiac transplantation. The right rectus sheath is entered in the midline and the right rectus abdominis muscle elevated from the posterior rectus fascia (Fig. 6.5). The medial aspect of the right hemidiaphragm and adjacent transversus thoracis muscles are taken down from the right side of the sternum, xiphoid and costal arch. If necessary, the right pleural space is opened and the outlet graft positioned medial to the right lung. With the heart decompressed it is usually possible to close the pericardium even in the presence of VAD cannulae. Pay close attention to VAD filling and VAD flow as pericardial closure is completed. If there is a decline in VAD filling or flow the pericardium is left open. Pericardial closure is facilitated by using a pericardial membrane (Preclude pericardial membrane; W.L. Gore and Associates, Inc., Flagstaff, AZ).[6] The membrane is sewn to the lateral edges of the pericardium at least to the level of the xiphoid process. This will further protect the right heart and outlet graft at the time of sternal reentry. The usual mediastinal tubes are employed; one anterior to the heart and one along the diaphragmatic surface. If an implantable blood pump is located in a preperitoneal position the pocket is drained with two 10 mm flat, fluted silicone drains (Blake drain; Johnson & Johnson Medical, Inc., Arlington, TX), one of which is located ventral and one dorsal to the blood pump. These drains are exteriorized through the dependent portion of the preperitoneal pocket. The sternum and midline incisions are closed as they would be for any routine open heart operation.

POSTOPERATIVE MANAGEMENT

IN-HOSPITAL MANAGEMENT CONSIDERATIONS

As the patient is moved from the operating room to the intensive care unit one individual is assigned to monitor the VAD drive unit. At our institution a perfusionist is responsible for safe transportation of the drive unit and monitoring of VAD filling and flow. In general, the patients are in an automatic pumping mode at the time of transfer. Regardless of device employed, the automatic mode ensures that the cardiac output is maximized. The fundamental function of the VAD in the automatic mode is that the blood pump enters systole each time it fills. As the VAD spends little time at end-diastole, systemic perfusion is optimized. The perfusionist is required to remain in attendance with the patient for the first 12-24 hours. When the patient's condition stabilizes the perfusionists provide coverage from outside the hospital.

The first order of business on arrival in the intensive care unit is to correct the coagulopathy if one is present. As is the case in the operating room blood component therapy is directed by a specific coagulation abnormality. The goal is a hematocrit of 30% with normal coagulation parameters although blood product administration is minimized to avoid antigen exposure. Mediastinal tube drainage is carefully monitored. Anticoagulation is not initiated until mediastinal tube drainage falls below 50-100 ml/hour for at least 6 hours. Anticoagulation is usually withheld for the first 12-24 hours. For the patient receiving mechanical ventricular assistance with a Thoratec VAD, a continuous heparin sodium infusion is begun when mediastinal tube drainage slows. The ACT is maintained at approximately 200 seconds. When the patient is able to take oral medication warfarin sodium is begun. The continuous heparin sodium infusion is discontinued when the international normalized ratio (INR) exceeds 2.5. The goal for anticoagulation in the Thoratec patient is an INR between 2.5 and 3.5. A similar approach is recommended for the patient requiring mechanical circulatory support with the Novacor VAD. For patients receiving mechanical ventricular assistance with the Thermo Cardiosystems VAD, aspirin (325 mg) is administered either by suppository or orally beginning on postoperative day one.

Patients with end stage heart failure have an excess of total body water. Intake is minimized by concentrating all intravenous infusions and eliminating maintenance intravenous fluids. As the systemic perfusion is determined by VAD flow there is no need for frequent cardiac output determinations. A thermodilution cardiac output may be obtained once or twice daily to ensure that it correlates with the VAD flow reported by the drive console. A continuous furosemide infusion ensures a forced diuresis during the first 24-48 hours following VAD implantation. One hundred milligrams of furosemide are mixed in 100 ml of 5% dextrose solution. The infusion is started at 10 ml/hour and increased while carefully monitoring intake and output. The goal is for total output to exceed intake by 1-2 liters over the first 24 hours. The furosemide infusion is titrated to achieve this endpoint. With aggressive diuresis the patient will develop a metabolic alkalosis. The metabolic alkalosis is corrected with acetazolamide sodium (250-500 mg IV q6h). Correction of the metabolic alkalosis permits early extubation. Thereafter, patients are instructed in the use of incentive spirometry. Respiratory treatments are employed if indicated. Chest physical therapy is gentle, if necessary, but rarely required.

Infection control is a critical issue in a patient who will ultimately undergo cardiac transplantation with the attendant immunosuppression. All tubes, lines and the urinary drainage catheter are removed as soon as possible. The mediastinal tubes are generally withdrawn within 12-18 hours. The drains within a preperitoneal pocket for the intracorporeal ventricular assist pumps remain in place for a longer period of time. There is little bleeding from the preperitoneal pocket; however, the extensive dissection usually results in a significant serous fluid leak. If these drains are removed too early the patient will develop a pocket seroma which can become secondarily infected. Alternatively, the serous fluid may track along the percutaneous drive line and exit the wound at the skin level. Such a serous fluid leak makes it difficult for the skin to adhere to the textured surface of the percutaneous drive line. Satisfactory tissue ingrowth at this skin:biomaterial interface is the best defense against an ascending drive line infection, one of the more prominent sources of long-term morbidity in the patient who requires an extended period of mechanical circulatory support. The preperitoneal pocket drains are placed to -125 mm Hg suction (Varidyne canister kit; Surgidyne, Inc., Minneapolis, MN). The drains remain in place until the total drainage is less than 50 ml over a 24 hour period. These drains routinely remain in place for 1-2 weeks. There is no need for prolonged antibiotic coverage. In fact, ongoing antibiotic administration is counterproductive in that it may permit overgrowth of resistant species.

Rehabilitation begins on the first postoperative day.[7] The physical therapists evaluate the patient and range of motion exercises begin within 12-24 hours. The patients are generally out of bed and able to ambulate a short distance within 24-48 hours. Nutritional supplementation begins immediately. Patients receive a feeding tube within the first 24 hours and enteral supplementation is initiated as soon as peristalsis returns. If an implantable VAD is placed within the abdomen a postoperative ileus may delay initiation of tube feeds. In this instance, it may be wise to resort to central venous hyperalimentation although it is preferable to avoid that route due to obvious infectious concerns.

Patients are usually ready to be transferred out of the intensive care unit within 2-4 days. The most common cause for a delay in transfer out of the intensive care unit is persistent right ventricular dysfunction. In most circumstances, the fall in pulmonary artery pressures associated with left ventricular assistance is accompanied by slow improvement in right ventricular contractility over the first few postoperative days. As right ventricular function improves inotropic therapy and pulmonary vasodilators are withdrawn. If the patient is supported with an RVAD, the return of right ventricular function is documented by converting the RVAD drive unit to a fixed rate mode and slowly decreasing the RVAD pumping rate. The central venous pressure and LVAD filling and flow are carefully monitored as RVAD flow is reduced. If the patient is able to maintain a central venous pressure below 15 mm Hg and an LVAD flow index in excess of 2.0 L/min/m^2, the RVAD can be safely removed. In the absence of systemic anticoagulation RVAD flow should not be reduced below 2.0 L/min. If it appears that the patient's right ventricular function has improved the patient is returned to the operating room for RVAD explantation. CPB is not employed. Systemic anticoagulation is unnecessary. The RVAD atrial cannula is withdrawn and the pursestring sutures tied. The outlet graft to the main pulmonary artery is transected and the stump of the graft adjacent to the main pulmonary artery oversewn. At cardiac transplantation the main pulmonary artery is transected distal to the outlet graft to remove the graft remnant. The RVAD inlet and outlet cannulae are withdrawn from the subcostal region and the fascial planes debrided and closed with absorbable sutures. The skin sites are also debrided and closed primarily.

On transfer out of the intensive care unit the patient is moved to a private room on the general ward.[8] The room should be large enough to house the drive unit, hardware and additional supplies for both the VAD and drive line or cannulae dressing changes. In order to create a home-like environment visitation rules are suspended. The patient's family and spouse are offered unlimited visitation and arrangements made for overnight spousal accommodation. When medically feasible the frequency with which vital signs are determined is decreased and nighttime vital sign checks eliminated altogether.

As the patient moves out of the intensive care unit, rehabilitation efforts assume primary importance. Dietary consultation is obtained. Initially tube feedings are continued around the clock and the patient's caloric intake documented. When the patient is ready for an oral diet, the tube feedings are cycled to nighttime only with the feeding tube clamped during the day. The patient is offered food that he enjoys utilizing a sodium restriction only as necessary. In the immediate postoperative period it is not necessary to pursue either a low fat or low cholesterol diet. When the patient's appetite improves, as documented by calorie counts, a heart healthy diet is instituted. When the patient's oral intake meets his caloric needs the feeding tube is removed.

Cardiac rehabilitation should be equally aggressive as patients with end stage heart failure are severely deconditioned. A formal graded exercise program is initiated under the direction of a trained physical therapist. The patient begins with weight bearing at the bedside followed by ambulation with assistance. Use of a treadmill in a monitored environment ensures that the patient's rehabilitation is progressive. It is important to note that although the VAD drive unit is responsive to the patient's physiologic need when functioning in the automatic mode there is a slight delay, and thus, the patient can experience orthostatic hypotension. Instruct the patient to sit at the edge of the bed before rising as there is a finite response time for the increase in VAD flow when the patient changes position.

Ultimately, the goal is to provide the patient with as much independence as possible. The patient is instructed as to the function of the VAD and the patient and other appropriate family members are taught to respond to alarm situations.

The patient can usually be removed from the cardiac monitor within a week or two. This allows ever increasing independence as the patient may leave the general ward initially in the company of a caregiver and ultimately with another family member. During this time the patient is given increasing responsibility for his own medical care. The patient is provided with and instructed to take his own medications. A family member or other caregiver is taught to perform cannulae or drive line dressing changes. Using sterile technique the drive line or cannulae exit sites are irrigated with hydrogen peroxide (3% diluted 50:50 with normal saline) and covered with a thin layer of 1% silver sulfadiazine cream and a sterile dressing. Initially the caregiver performs the dressing changes under the supervision of hospital personnel. Ultimately, the caregiver is permitted to perform the dressing changes independently. Infection control is ongoing. In addition to fastidious cannulae and drive line exit site care, the patient is taught to avoid motion of the cannulae/drive line at the cannulae or drive line-skin interface. This ensures tissue ingrowth into the textured surface of the percutaneous cannulae or drive line.

HOME DISCHARGE

Currently, patients supported with the Thoratec VAD or Thermo Cardiosystems 1000 IP (pneumatic) LVAS must remain in the hospital during the period of VAD support. Patients supported with either the Thermo Cardiosystems VE (electric) LVAS or Novacor N-100 LVAD are candidates for home discharge (see Chapter 9: Nursing Care of the Patient Requiring Mechanical Circulatory Support).[9] Prior to discharge from the hospital the patient and significant other must be able to manage the VAD and drive unit and respond appropriately to common alarm situations, including device failure. The patient and family must be comfortable in performing drive line dressing changes. Arrangements are made to ensure that the patient continues in a cardiac rehabilitation program at a local healthcare facility. Patients are not permitted to drive an automobile. Although they are encouraged to continue moderate aerobic activities we ask that they avoid high impact sports such as running and basketball. To ensure a seamless transition at the time of hospital discharge our transplant/VAD coordinators travel to the patient's home community to provide educational programs to local healthcare providers. In particular, the coordinators meet with the local emergency medical service and cardiac rehabilitation personnel. Close contact is maintained with the patients referring internist/cardiologist. Most medical emergencies can be handled through the local healthcare facility. In the event of device failure the patient should be transported directly to the center at which the VAD was implanted. To ensure that postoperative teaching is complete we ask that the patient and significant other spend the first night following discharge from the hospital in the immediate vicinity of the implant center. The patient returns to the outpatient clinic for a final check the following morning. Such a "dry run" frequently identifies educational deficiencies helping to ensure that the patient and family caregiver(s) are adequately prepared and comfortable with the return home. Patients who are discharged to home are required to return to the outpatient clinic every 2-4 weeks for the duration of VAD support.

PREPARATION FOR CARDIAC TRANSPLANTATION

When the patient is taken to the operating room for mechanical blood pump insertion, he is deactivated from the cardiac transplant waiting list. Prior to VAD insertion the patient is classified as a UNOS Status 1A. Postoperatively the patient remains UNOS Status 7 until such time as the patient's recovery is complete and he is prepared to undergo a cardiac transplant. Frequent wound checks and inspection of the percutaneous cannulae or drive line exit site(s) ensures that the patient's incisions have healed adequately prior to relisting. The prerenal azotemia and hepatic dysfunction related to systemic hypoperfusion prior to VAD insertion take several weeks to resolve.[10] Initially, a serum BUN and creatinine are obtained on a daily basis. As the patient convalesces the frequency of laboratory analysis is decreased. Renal and hepatic profiles are obtained on a weekly basis until the values normalize. Thereafter, monthly laboratory analyses of renal and hepatic function usually suffices. A serum total protein, albumin and prealbumin document the patient's nutritional status. Transplantation is not performed until the patient achieves positive nitrogen balance and the serum prealbumin normalizes. A percent reactive antibody (PRA) is also determined on a weekly basis to ensure that the patient has not developed systemic antibodies related to a perioperative blood transfusion. The weekly PRA determination continues for the duration of mechanical circulatory support. If the patient develops a positive PRA a prospective crossmatch is required prior to cardiac transplantation. On average, patients remain classified as UNOS Status 7 for 2-4 weeks. Barring unforeseen circumstances, most patients are returned to an active UNOS status within 1 month of mechanical blood pump insertion. The recent change in the definition of the UNOS classifications dictates that patients who have had a mechanical blood pump in place in for less than 30 days be listed as a Status 1A. Patients who have received mechanical circulatory support for 30 days or more are listed as UNOS Status 1B. Patients supported with a VAD for more than 30 days who develop a complication related to the mechanical blood pump (infection, stroke, device malfunction) are eligible to be reclassified as UNOS Status 1A.

Patients with preexisting pulmonary hypertension are taken to the cardiac catheterization laboratory for a right heart catheterization every 2-3 months. Interval determination of pulmonary artery pressures documents the rate at which these pressures fall during a prolonged period of left heart support. The presence and severity of pulmonary hypertension determines, in part, donor selection. A patient with pulmonary hypertension is best served by a donor heart that is oversized (donor:recipient weight ratio > 1) with a short ischemic time. As pulmonary artery pressures fall in the patient receiving LVAD support, donor heart criteria are liberalized. It may be possible to accept an undersized heart (donor:recipient weight ratio < 1) with a longer ischemic time as pulmonary artery pressures normalize.

During the period of mechanical circulatory support the patient is educated about issues pertinent to cardiac transplantation. The patient is taught to take his own vital signs and is educated about the medications that will be prescribed following cardiac transplantation. The patient and the patient's family are visited by a social worker, clinical psychologist or psychiatrist. These patients have been hospital bound with a chronic illness for an extended period of time. There are a multitude of issues that can impact a patient's sense of well-being including loss of income, mounting hospital bills and loss of family unity. The patient's coping skills can be further impaired if complications develop or other setbacks occur.

CARDIAC TRANSPLANTATION

When a donor heart becomes available the patient is moved from the general ward to the surgical intensive care unit. If the patient has been discharged to home the patient returns to the hospital and is admitted directly to the surgical intensive care unit. A radial arterial line and pulmonary arterial catheter are inserted. Femoral arterial catheters are avoided due to the increased risk of line sepsis. If a groin catheter is required the right common femoral artery and vein are preserved for cannulation for CPB. The patient is clipped from chin to knees, bedside to bedside. If the patient has a positive PRA a prospective crossmatch is performed. Perioperative antibiotic coverage consists of vancomycin hydrochloride (1 gm IV) and cefotaxime sodium (1 gm IV). Full dose aprotinin is employed. As the patient received aprotinin at the time of VAD insertion, the pulmonary artery and systemic pressures are carefully monitored during test dose administration. If pulmonary artery pressures rise precipitously during test dose administration aprotinin administration is discontinued. Four units of leukodepleted packed red blood cells are made available. We have not found it necessary to resort to the use of a separate blood scavenging system (Cell Saver Pac; Baxter Healthcare Corp., Irvine, CA). The services of two perfusionists are required for VAD explantation and cardiac transplantation. One perfusionist runs the CPB machine while the second perfusionist is responsible for the VAD drive console.

The skin prep is modified to isolate the percutaneous skin sites from the remainder of the operative field. This is easily accomplished with an implantable device as the percutaneous drive line exit site is located a distance from the sternum. In this situation, an assistant elevates the drive line while the standard skin prep is performed. The percutaneous drive line exit site is prepped as is the adjacent 20 cm of drive line. The patient is draped in a standard fashion. The drive line is wrapped in sterile towels and laid toward the side of the operative field. The skin exit site itself is covered with a separate towel. The entire field is then covered with a sterile self-adhesive drape (Ioban drape; 3M Health Care, St. Paul, MN). The adhesive is not placed directly on the drive line or skin exit site. The presence of the sterile towel between the adhesive drape and drive line allows the cardiac transplant to be performed in an operative field that does not include a potentially colonized skin drive line exit site. Furthermore, upon completion of the cardiac transplant, prior to removing the sterile drapes, the towel covering the skin exit site is lifted and the drive line withdrawn from beneath the operative drape. The exit site is debrided and closed following closure of the sternotomy.

If the Thoratec VAD is employed a similar draping technique is utilized. An assistant gently elevates the Thoratec VAD taking care to avoid bending either the inlet or outlet cannulae. The skin prep includes the adjacent segment of the inlet and outlet cannulae. The patient is draped and separate sterile towels are wrapped about the Thoratec VAD. The cannula exit sites are covered with a separate sterile towel. By so doing, the cannulae exit sites are isolated from the sternotomy but easily accessible to allow VAD removal following recipient cardiectomy.

Timing is an issue as the amount of dissection required for recipient cardiectomy and device explantation, in particular, with an implantable device can be extensive.[11] If the patient is supported with an implantable device we allow 2 hours from skin incision to reach the point at which we are prepared to begin implanting the donor heart. Thus, it may be necessary to begin operating on the recipient before the donor heart has been removed. The recipient operation begins by opening the right groin. The common femoral artery and vein are isolated. The right groin is preferred as the inferior vena cava is readily cannulated with a long venous cannula (Femoral venous cannula and introducer set; Medtronic Bio-Medicus, Inc., Eden Prairie, MN). The left common femoral vein is avoided as passage of a long venous cannula over the sacral promontory may be problematic. CPB lines are brought to the operative field and one limb of the "Y" in the venous line is extended with a three-eighths inch ID tubing to allow cannulation of the superior vena cava within the chest and inferior vena caval cannulation in the groin. Heparin sodium is drawn up and immediately available to the anesthesiologist while appropriately typed, crossed and leukodepleted blood is present in the room at the time of sternotomy.

The sternotomy is performed keeping in mind that regardless of VAD employed the outlet graft may lay immediately beneath the sternum (Fig. 6.9). If the pericardium has been closed with a pericardial membrane, there is usually a clean plane of dissection beneath the sternum. Complete as much dissection within the mediastinum as possible prior to heparinization in order to decrease postoperative bleeding. The inlet conduit is not exposed during the initial dissection to avoid air embolism. If the patient has been supported with a Thoratec VAD, diastolic vacuum is discontinued prior to opening the sternum. In the case of an implantable VAD, the inlet conduit is not exposed until CPB is initiated and the outlet graft clamped. Random cultures are obtained from the mediastinum and the preperitoneal pocket in the case of an implantable VAD.

The patient is fully heparinized and the ACT measured in excess of 480 seconds prior to initiating CPB. Bicaval venous cannulation is employed. The superior vena cava is cannulated directly while the inferior vena cava is adequately drained using the long femoral venous cannula. To preserve aortic length, arterial cannulation is accomplished via the common femoral artery. As CPB is initiated VAD pumping is immediately discontinued and a vascular clamp applied to the VAD outlet graft. The aorta is crossclamped as far distally as possible and the

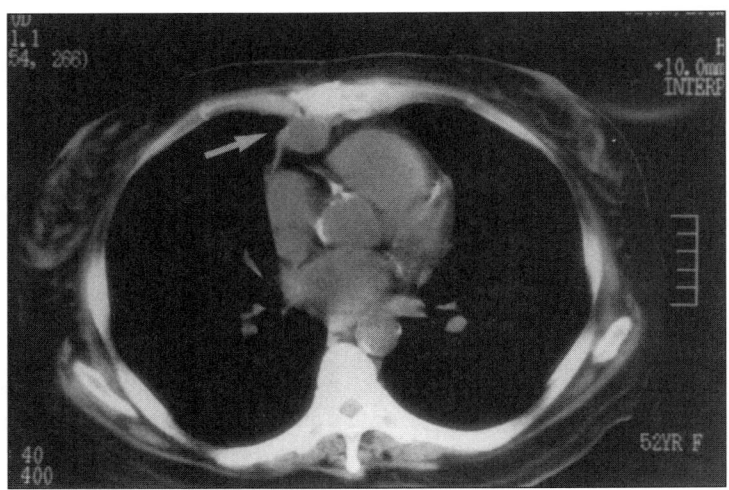

Fig. 6.9. Computed tomography of a patient supported with the Thermo Cardiosystems VE LVAS. The outlet graft (arrow) is located beneath the sternum just to the right of the midline.

superior and inferior vena cavae snared with umbilical tapes. Once the heart has been isolated from the systemic circulation VAD explantation occurs. If the patient is supported with a Thoratec LVAD using a left atrial inflow cannula, the inflow and outflow cannulae are transected with heavy scissors at the caudad pericardial reflection. The paracorporeal blood pump and attached cannulae "stumps" are withdrawn from beneath the operative drapes. The remnants of the left atrial and aortic cannulae are removed from the mediastinum. The left atrial cannula is withdrawn and the surrounding pursestring sutures tied. Never withdraw the left atrial inflow cannula before the aortic crossclamp is applied. Once the recipient cardiectomy is complete carefully inspect the inside of the recipient left atrium to ensure that all thrombus/fibrin which collected around the left atrial cannula entry point has been removed. Both maneuvers reduce the potential for an intraoperative embolic event. The aortic outlet graft stump is removed by transecting the recipient aorta just beyond the outlet graft-to-aortic anastomosis (Fig. 6.10). The amount of biomaterial remaining in the chest following cardiac transplantation is thereby reduced. If the patient has been supported with a Thoratec LVAD using a left ventricular apex inflow cannula, the inlet cannula must be dissected away from the diaphragm before the cannula can be transected and the blood pump removed from beneath the sterile drapes. Adhesions at the left ventricular apex can be particularly dense. To facilitate exposure the diaphragm may be taken down from the chest wall. An alternative, less morbid technique is to simply transect the heart in the mid portion of the left ventricle and remove the left ventricular apex in a piecemeal fashion.

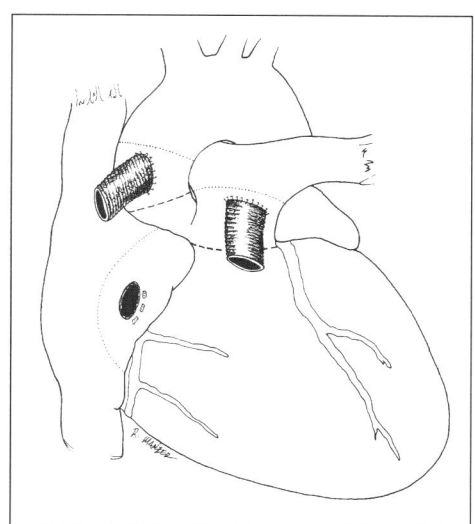

Fig. 6.10. Modified recipient cardiectomy. The great vessels are normally transected just distal to the semilunar valves (dashed lines). To remove the outlet graft stumps, the aorta and pulmonary artery are transected just distal to the graft-to-great vessel anastomoses (dotted lines). The right atrial cannulation site is removed along with the adjacent wall of the right atrium (dotted line).

6

If the patient has been supported with an implantable LVAD, the drive line is transected adjacent to the body of the blood pump allowing the drive line to be withdrawn from the percutaneous tunnel beneath the operative drapes by a nonsterile assistant. The left ventricular apex inflow and aortic outflow conduits are removed in a manner identical to that described for the Thoratec LVAD above. The implantable blood pump is withdrawn from the pre- or intraperitoneal pocket as the inlet and outlet cannulae are removed. Regardless of VAD(s) employed, the percutaneous skin cannulae or drive line exit site(s) are kept isolated from the sternotomy to avoid cross contamination. The skin exit sites are not exposed until the cardiac transplant is completed and the chest closed.

If the patient also received right ventricular assistance, the RVAD inlet cannula is withdrawn from the right atrium and the right atrial free wall debrided following recipient cardiectomy. All pursestring suture material is removed leaving pristine right atrial tissue behind. The main pulmonary artery is transected just distal to the RVAD outlet graft-to-pulmonary artery anastomosis. We utilize the standard Shumway cardiac transplantation technique. Thus, the recipient cardiectomy is completed by dividing the atria adjacent to the atrioventricular groove. The sole exception is the additional right atrial free wall debridement in the patient who has required right ventricular assistance.

Time permitting, the pericardium and preperitoneal pocket, if present, are debrided and irrigated. The diaphragmatic defects created by the cannulae are closed with absorbable sutures. The donor heart is implanted in a manner identical to that utilized in a patient that did not require mechanical ventricular assistance. The sole difference in donor management is the donor cardiectomy which is modified to remove as much aortic length as possible. If the recipient has required right ventricular assistance and the main pulmonary artery remnant is shorter than normal, the donor cardiectomy is also modified to include as much main pulmonary artery length as possible.

Once CPB has been discontinued and preparations are made to close the chest, a separate left pleural tube is inserted in addition to the usual mediastinal tubes. Removal of a left ventricular apex cannula usually results in an open left pleural space. If the patient received mechanical circulatory support with an implantable device, the pseudocapsule is removed from the preperitoneal pocket. The drive line exit site is closed from the inside with absorbable sutures. The preperitoneal pocket is drained with soft Silastic drains and the dead space obliterated by closing the pocket with multiple interrupted absorbable sutures. Elimination of dead space and adequate drainage decreases the potential for subsequent pocket seroma development. The sternotomy incision is closed and dressed prior to exposing the drive line or cannulae skin exit sites. These skin sites are debrided by ellipsing the skin and underlying soft tissue. All necrotic or infected tissue is carefully removed from the subcutaneous tunnel. The underlying fascia is closed with absorbable sutures. If the subcutaneous tunnel is obviously infected, the skin and soft tissues are packed open. If the drive line or cannula tract does not appear to be colonized the subcutaneous tunnels can be closed primarily. Keep in mind that posttransplant steroid administration delays wound healing. Thus, if the drive line/cannulae sites are packed open it will take additional time for these areas to granulate closed.

POSTTRANSPLANT CARE AND MANAGEMENT

Posttransplant care is identical to that provided for a patient who has undergone cardiac transplantation without a pretransplant period of mechanical circulatory support. Patients who required mechanical ventricular assistance prior to transplantation enter this operation fully rehabilitated with normal end organ function. Thus, recovery following VAD explantation and cardiac transplantation is usually quite rapid. As the patients are adequately prepared for their transplant both from a physiologic and educational standpoint, preparations are made for an early hospital discharge.

Management considerations peculiar to patients requiring mechanical circulatory support prior to cardiac transplantation are primarily related to wound management. If the patient has a colonized drive line or cannula exit site at the time of cardiac transplantation, we employ routine perioperative antibiotics with no need for ongoing antibiotic therapy. If, however, the patient demonstrates systemic sepsis related to ventricular assist pump colonization, at the time of the cardiac transplant antibiotics should be continued for 4-6 weeks following cardiac transplantation. For this reason it is helpful to disassemble the VAD at the time of cardiac transplantation in order to identify and culture any vegetations, if present.

If the drive line/cannulae exit sites have been closed primarily, they can be treated as any other wound. If an implantable blood pump was employed the drains remain in the preperitoneal pocket until drainage falls below 50 ml in a 24 hour period. These drains are connected to a Varidyne canister at -125 mm Hg suction in the interim. If the drive line/cannulae exit sites have been packed open the dressing should be changed twice daily using saline wet-to-dry dressing sponges. These sites will usually granulate closed over a period of 1-2 months.

REFERENCES

1. Frazier OH, Rose EA, McCarthy P et al. Improved mortality and rehabilitation of transplant candidates treated with a long-term implantable left ventricular assist system. Ann Surg 1995; 222:327-338.
2. Pennington DG, McBride LR, Peigh PS et al. Eight years' experience with bridging to cardiac transplantation. J Thorac Cardiovasc Surg 1994; 107:472-481.
3. Thoratec's Heartbeat Newsletter. 1998; 12:2.
4. Koul B, Solem JO, Steen S et al. HeartMate left ventricular assist device as bridge to heart transplantation. Ann Thorac Surg 1998; 65:1625-1630.
5. Farrar DJ, Hill JD, Pennington DG et al. Preoperative and postoperative comparison of patients with univentricular and biventricular support with the Thoratec ventricular assist device as a bridge to cardiac transplantation. J Thorac Cardiovasc Surg 1997; 113:202-209.
6. Holman WL, Bourge RC, Zorn GL et al. Use of expanded polytetrafluoroethylene pericardial substitute with ventricular assist devices. Ann Thorac Surg 1993; 55:181-183.
7. Mancini D, Goldsmith R, Levin H et al. Comparison of exercise performance in patients with chronic severe heart failure versus left ventricular assist device. Circulation 1998; 98:1178-1183.
8. Reedy JE. Transfer of a patient with a ventricular assist device to a noncritical care area. Heart Lung 1993; 22:71-76.
9. Pristas JM, Winowich S, Nastala CJ et al. Protocol for releasing Novacor left ventricular assist system patients out-of-hospital. ASAIO J 1995: 41:M539-M543.
10. Burnett CM, Duncan JM, Frazier OH et al. Improved multiorgan function after prolonged univentricular support. Ann Thorac Surg 1993; 55:65-71.
11. Oz MC, Levin HR, Rose EA. Technique for removal of left ventricular assist devices. Ann Thorac Surg 1994; 58:257-258.

6

Special Situations

Wayne E. Richenbacher

INTRODUCTION

The purpose of this chapter is to address a number of programmatic and patient management issues that do not conveniently fit into the other chapters in this book. The comments in the section on interhospital transfer are primarily directed to the patient with postcardiotomy cardiogenic shock who fails to recover ventricular function and must be transferred from a nontransplant center to a hospital that performs cardiac transplantation. The remarks in this section are also referable to the management of the bridge patient who is cared for in a community located at a distance from the transplant hospital. If the patient is supported with an implantable blood pump and is living at home awaiting a cardiac transplant, device or nondevice related problems may necessitate transfer of the patient back to the implanting facility. Ventricular arrhythmias or native valvular disease may occur in patients requiring mechanical circulatory support either as a bridge to cardiac transplantation or for postcardiotomy cardiogenic shock. The management recommendations for patients with either of these two conditions serve to complement the patient care chapters in this text. Primary right ventricular failure rarely occurs in the bridge or postcardiotomy cardiogenic shock scenario. It is a unique circumstance that warrants comment as many management issues are peculiar to this clinical condition.

INTERHOSPITAL TRANSFER

Interhospital transfer is recommended for patients who require mechanical circulatory support for postcardiotomy cardiogenic shock, who are cared for at a nontransplant facility and who demonstrate no evidence of ventricular recovery. If a cursory examination reveals no major contraindication to cardiac transplantation, the patient should be referred to a transplant center for definitive management of his heart failure (Table 5.1). The transfer must be carefully planned and coordinated to ensure that the transition is seamless in the critically ill patient. Local/regional transfer is best accomplished via an ambulance. Documentation to accompany the patient on transfer is listed in Table 7.1. The discharge summary should describe the patient's primary medical problem and include copies of preoperative diagnostic studies, particularly those that provide an estimate of ventricular function. The postoperative course should be summarized with particular attention directed to events that may impact the patient's candidacy for cardiac transplantation. These would include, but are not limited to, neurologic sequelae, renal and hepatic function and septic complications. An operative note describing the details of the original open heart operation as well as the mechanical blood pump implantation is most helpful when planning another operation to convert the patient to a long-term ventricular assist device (VAD). A blood transfusion record including blood products utilized and whether or not they were leukodepleted gives some indication of the antigen load to which the patient has been exposed. Studies used to assess myocardial function following VAD implantation including serum creatine kinase or troponin I levels provide an estimate of the extent of the perioperative myocardial injury.

7

At the time of transfer, personnel from the transplant center should travel to the referring hospital to assess the patient and aid in transfer. During transfer the requisite caregivers include a nurse, second nurse or perfusionist to assume responsibility for the VAD drive unit and a respiratory therapist if the patient is intubated. The ambulance should be of sufficient size to permit transportation of the patient, the VAD drive console, ventilator, nurses and transport personnel. Given the highly specialized hardware and physiology associated with the patient requiring mechanical circulatory support, it is best that the ambulance crew be assigned to provide access to supplies and assist the primary caregivers in this instance. Check the power requirements of the equipment to be transported. The power source available should be capable of running one or two mechanical blood pump consoles, monitors, intravenous pumps and a mechanical ventilator. Back-up hand pumps or alternate power source for the VAD should be available and easily accessed in the event of a power failure. Additional medications including

---

Table 7.1. Records to accompany the mechanical blood pump patient on transfer from one healthcare facility to another

Discharge summary
Preoperative studies
    Myocardial function (left ventriculogram, radionuclide ventriculogram)
    Right heart catheterization (reversibility of pulmonary hypertension, if present)
Operative note
    Details of original open heart operation
    Details of ventricular assist device implantation
Studies used to assess perioperative myocardial injury
    Serum creatine kinase, troponin I levels
Blood transfusion record
Studies used to assess myocardial function after ventricular assist device implantation

inotropes are readily available. During transportation the patient's hemodynamics and VAD function are carefully monitored. The patient is admitted directly to the surgical intensive care unit upon arrival at the tertiary care center.

Transportation of a patient who is supported with an Abiomed BVS 5000 Ventricular Assist System (Abiomed, Inc., Danvers, MA) warrants special comment. As this device is gravity filled, VAD flow(s) may be suboptimal when the patient is placed on an ambulance stretcher. These stretchers are located quite near the floor of the ambulance which may make it difficult to keep the blood pump below the level of the patient. If VAD filling is problematic prior to transportation it may be safest to convert the patient to a centrifugal pump for the actual transfer. The advantages of such a conversion include avoidance of the problem with gravity filling and the need to keep the pump below the level of the patient within the close confines of the ambulance. Secondly, the centrifugal pump console is considerably smaller than the Abiomed drive console.

Transportation of the mechanical blood pump patient to a distant center may require the use of an aircraft. Unfortunately, current experience with air transfer of the VAD patient is anecdotal and not nearly as well defined as air transportation of the patient requiring intraaortic balloon counterpulsation. In general, a larger aircraft is required than that used to transport a patient with an intraaortic balloon pump (IABP). If a larger fixed wing or rotor winged aircraft is employed two pilots may be necessary and landing requirements may differ. The latter regulations are determined by the Federal Aviation Administration. The personnel required for patient care during transportation on an aircraft varies with the healthcare institution. There are no federal regulations in this regard. At The University of Iowa it is mandated that a pilot and two nurses accompany a patient during air transportation. Another consideration pertinent to air transportation is the change in air pressure with altitude. The IABP is a closed system subject to Boyle's Law. Mechanical blood pumps vary in design and may or may not be subject to Boyle's Law. Thus, it is important to ascertain ahead of time whether or not the particular device to be transported has undergone pressurization studies and is approved to fly in pressurized or nonpressurized aircraft. In general, pressurized aircraft have an internal pressure equivalent to 5000-8000 feet of elevation. Keep in mind, however, the potential for accidental loss of cabin pressure which may have a profound impact on a VAD or IABP that employs a closed system. It is also important to note that free air within body cavities will tend to expand by 25% during air travel. Thus, air travel is contraindicated in patients with trapped air such as those who have had a pneumothorax or a patient who has undergone a laparotomy, as occurs with intraperitoneal placement of an implantable VAD, within the previous 10 days. The issue of air travel by the patient requiring mechanical circulatory support will become more important in the future as portable drive units are approved and home discharge programs become more widespread. Air transportation will become less of an issue for the acute patient who requires transfer from one hospital to another but will become a quality of life issue in a patient who has received mechanical circulatory support either as a bridge to cardiac transplantation or as a permanent device implantation.

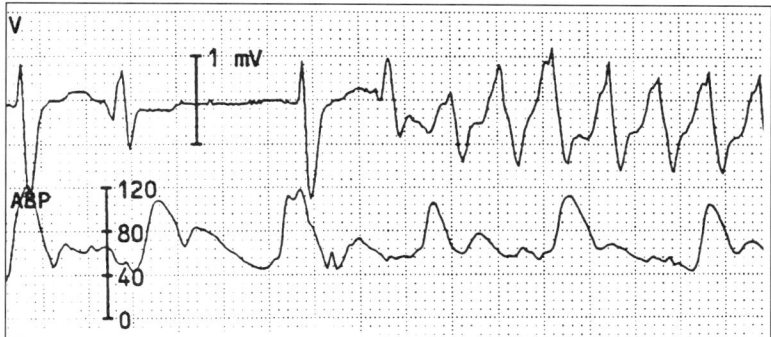

Fig. 7.1. Simultaneous electrocardiogram (V) and arterial blood pressure (ABP) traces in a patient supported with a left ventricular assist device. When the patient is in sinus rhythm there is asynchrony between the electrocardiogram and the pulse in the arterial pressure trace. The patient is device dependent; there is no native ejection. When the patient develops ventricular tachycardia, systemic perfusion is maintained. The ventricular assist device continues to pump blood despite the ineffective ventricular rhythm.

VENTRICULAR ARRHYTHMIAS AND ELECTRICAL DEFIBRILLATION

Ventricular tachyarrhythmias, defined as ventricular tachycardia or ventricular fibrillation, occur frequently in cardiomyopathic patients. These arrhythmias can persist following VAD insertion. Ventricular arrhythmias are well tolerated by the VAD patient while most arrhythmic episodes are brief in duration and resolve spontaneously (Fig. 7.1). It was originally thought that the presence of malignant ventricular arrhythmias was an indication for biventricular support. However, it has been shown repeatedly that patients supported with a left VAD (LVAD) alone are tolerant of ventricular arrhythmias.[1] VAD flow may fall 1-2 L/min but the patients do not become syncopal or dyspneic. Even in the presence of ventricular fibrillation if a patient is supported with an LVAD and has a low pulmonary vascular resistance, flow through the right heart is maintained. Passive blood flow through the right heart is the hemodynamic equivalent of a Fontan (systemic vein to pulmonary arterial) circulation.

Patients who have suffered an acute myocardial infarction have a high incidence of ventricular arrhythmias following VAD insertion. Patients with nonischemic heart disease have a much lower incidence of ventricular arrhythmias following VAD insertion. Regardless of etiology ventricular arrhythmias usually resolve following VAD insertion. As hemodynamic decompensation rarely accompanies even a prolonged arrhythmic episode, time is permitted for pharmacologic intervention. In fact, the hemodynamic stability achieved by VAD insertion has led to the suggestion that refractory ventricular arrhythmias may serve as another potential indication for VAD support.[2]

Electrical cardioversion is reserved for patients who are refractory to medical therapy and is warranted to avoid thrombus formation in the native ventricle and improve exercise tolerance. Electrical cardioversion, even if performed on a patient in ventricular fibrillation, is always an elective procedure. The patient is appraised of the rationale for cardioversion and is sedated during the procedure. The Abiomed, Thoratec (Thoratec Laboratories Corp., Pleasanton, CA) and Novacor N-100 (Novacor Division, Baxter Healthcare Corp., Santa Cruz, CA) VADs are electrically isolated from the patient and require no special management during external defibrillation. The Thermo Cardiosystems HeartMate VAD (Thermo Cardiosystems, Inc., Woburn, MA), on the other hand, can be irreparably damaged by external cardioversion. The pneumatic version of the Thermo Cardiosystems VAD (1000 IP LVAS) should be converted to a fixed rate mode and the Hall sensor disconnected from the rear of the drive console prior to defibrillation. When defibrillation is complete the Hall sensor is reconnected to the pneumatic drive console and the drive unit converted back to the automatic mode. The electric version of the Thermo Cardiosystems VAD (VE LVAS) must be completely disconnected from the system controller and power base unit/battery packs prior to cardioversion. When defibrillation or cardioversion is complete, the VAD is reconnected to the system controller and power source. When the electric VAD is disconnected from the external hardware, VAD pumping ceases. If defibrillation is expected to take longer than 30 seconds, the VAD should be hand-pumped and consideration given to systemic anticoagulation. The same management protocol is followed when using intraoperative defibrillation. Even though a much lower energy shock is administered with internal paddles, the Thermo Cardiosystems electric VAD should be disconnected from the system controller and power source prior to defibrillation.

Patients who have been treated with an implantable cardioverter/defibrillator (ICD) warrant special consideration when being prepared for VAD insertion.[3] Originally, ICDs utilized epicardial sensing leads and epicardial or extrapericardial patch electrodes. The pulse generator was positioned in the subcutaneous tissue in the left subcostal region. The mere presence of the pulse generator in this location may make paracorporeal or implantable VAD insertion difficult. Care must be taken when positioning the Thoratec VAD inlet and outlet cannulae to avoid damaging the leads in the subcutaneous ICD pocket. An implantable VAD is positioned in a preperitoneal pocket with abdominal wall muscles separating the VAD pocket from the ICD pocket. The newer generation ICDs utilize transvenous sensing, pacing and defibrillator leads. The pulse generator is positioned along the chest wall well away from potential VAD cannulae skin exit sites or preperitoneal pocket.

During VAD implantation the ICD should be turned off to avoid inadvertent ICD discharge during the operative procedure. Once the VAD implant operation is complete a decision must be made regarding whether or not to reactivate the ICD. The VAD pumping action does not create a motion artifact that is recognized by the ICD arrhythmia sensing function. If the ICD discharges the low energy shock has no effect upon the electrical components of any VAD. As patients may continue to develop ventricular arrhythmias following VAD insertion, it is not unreasonable to leave the ICD on so that the arrhythmia will be terminated in a timely fashion. However, consideration must be given to the patient's sense of well-being. Patients who are supported with a VAD who develop a ventricular arrhythmia do not lose consciousness. The shock delivered by an ICD in a VAD patient can be an uncomfortable experience. We prefer to inactivate the ICD at the time of VAD insertion. If the patient develops a ventricular arrhythmia, as heralded by a fall in VAD flow or decrease in exercise tolerance in the nonmonitored patient, we employ pharmacologic cardioversion. In medically refractory situations external defibrillation is performed with appropriate intravenous sedation.

VALVULAR HEART DISEASE

Native valvular heart disease can have a profound impact on LVAD function and the ability of the mechanical blood pump to support the systemic circulation. If the patient has mitral stenosis, blood flow into the VAD is not compromised if left atrial cannulation is employed. If, on the other hand, left ventricular apex cannulation is utilized, the mitral stenosis must be corrected at the time of LVAD implantation. If severe mitral stenosis is not corrected there is an impediment to the flow of blood from the left atrium to the left ventricle. LVAD filling is impaired and LVAD flow decreased. As the left atrium is not adequately decompressed, left-sided filling pressures remain elevated and any underlying right ventricular dysfunction is exacerbated. If the mitral valve must be replaced valve prosthesis selection is based upon the LVAD employed and subsequent need for systemic anticoagulation.

When faced with a patient who has undergone mitral valve replacement, left atrial-to-aortic left ventricular assistance can be complicated by systemic thromboembolization even when a bioprosthesis is used in the mitral position and the patient is anticoagulated.[4] If the patient has a prosthetic valve in the mitral position LVAD inflow via the left ventricular apex ensures that the valve prosthesis will be well washed reducing the potential for valve thrombosis and valve failure.

Native mitral regurgitation is well tolerated regardless of whether left ventricular assistance employs left atrial or left ventricular inflow. It has been suggested that native left ventricular contraction, in the face of mitral insufficiency, can lead to left atrial hypertension. Maximal reduction in left atrial pressure is achieved by timing the LVAD to be countersynchronous with native ventricular contraction.[5] Although this timing sequence is difficult to achieve under the best of circumstances, left ventricular apex cannulation maximally decompresses the left ventricle ensuring effective left atrial decompression.

Native aortic stenosis, provided it is not associated with aortic insufficiency, is well tolerated by a patient requiring left ventricular assistance. If the LVAD employs left atrial inflow, the device functions in parallel with the native ventricle. If left ventricular inflow is employed, the LVAD functions in series with the native ventricle. In either instance outflow is into the ascending aorta downstream of the stenotic aortic valve. As left heart decompression is usually complete during left ventricular assistance, the native valve rarely opens when the LVAD is pumping. The only problem that may be encountered in the LVAD patient with aortic stenosis is in the event of a catastrophic device failure. If the LVAD were to suddenly stop pumping the patient may have inadequate left ventricular contractile reserve to force blood from the native ventricle through the stenotic aortic valve. This is only a problem with left atrial inflow. If left ventricular apex inflow is used the native ventricle can still eject through the LVAD itself.

Native aortic valve insufficiency is the most problematic valvular abnormality in the patient requiring mechanical left heart support. A preoperative echocardiogram is always performed prior to VAD insertion. The severity of aortic insufficiency is often underestimated on the preoperative echocardiogram as the patient with end stage heart failure is usually tachycardic and has systemic hypotension with a marked elevation in left ventricular end-diastolic pressure.[6] After LVAD insertion native aortic valve insufficiency increases as there is asynchrony between timing of LVAD systole and native left ventricular contraction, a low left ventricular end-diastolic pressure and the native aortic valve must remain competent continuously despite a normal LVAD generated aortic root pressure. If the native aortic valve is insufficient, blood ejected from the LVAD into the ascending aorta will travel back across the incompetent aortic valve into the left ventricle and reenter the VAD. The recirculation of blood through the LVAD reduces systemic blood flow and increases left sided filling pressures. If aortic insufficiency occurs late during the course of LVAD support, the condition is heralded by an inappropriate rise in LVAD flow. Echocardiography confirms the diagnosis.

If the aortic valve is replaced, the patient will be at risk for systemic thromboembolization for the duration of LVAD support. Most of the blood passes from the left side of the heart through the LVAD. As the aortic valve rarely opens, thrombus will form on a valve prosthesis even if the patient is fully anticoagulated. This is true regardless of whether a bioprosthesis or mechanical valve is employed. Thrombus that occurs on the valve can be ejected if the native heart contracts. The ideal solution to this clinical problem is unknown. Attempts to obliterate the left ventricular outflow tract by sewing the native aortic valve leaflets together at the time of LVAD insertion have been unsuccessful. The "repair" usually disrupts during the period of LVAD support. The most reasonable management approach for the patient with severe aortic insufficiency is to remove the native aortic valve and replace it with a bioprosthesis. Prior to implantation the bioprosthesis is modified by sewing an occlusive pericardial patch (Periguard pericardium, 4 x 4 cm; Bio-Vascular, Inc., Saint Paul, MN) over the left ventricular side of the valve (Fig. 7.2). The left ventricular outflow tract is thereby obliterated with a relatively nonthrombogenic well supported "valve prosthesis". Thrombus may still form on the aortic side of the prosthesis, thus long-term anticoagulation is appropriate. If the LVAD were to suddenly fail the native ventricle can still eject through the LVAD itself provided left ventricular apex cannulation was employed.

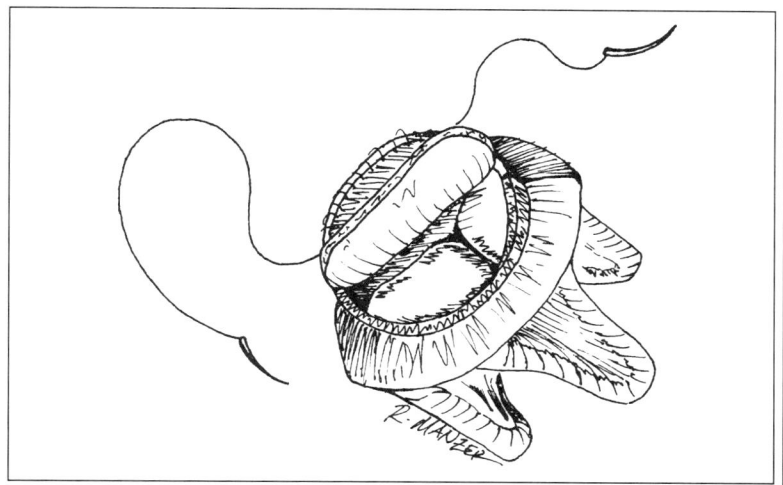

Fig. 7.2. The left ventricular outflow tract is obliterated when a left ventricular assist device is inserted in a patient with aortic insufficiency. An appropriately sized bioprosthetic valve is "modified" prior to insertion. A pericardial patch is sutured circumferentially to the sewing ring on the ventricular side of the aortic bioprosthesis. Thereafter the valve is implanted into the aortic position in the standard fashion

PRIMARY RIGHT VENTRICULAR FAILURE

Primary right ventricular failure is an unusual cause of postcardiotomy cardiogenic shock. The most common etiology is an acute occlusion of a dominant right coronary artery with a right ventricular myocardial infarction. Primary right ventricular failure may also occur in the setting of a pulmonary embolism complicated by right heart distension and decompensation. If the patient cannot be weaned from cardiopulmonary bypass (CPB) after coronary revascularization or pulmonary embolectomy, the hemodynamic and echocardiographic findings reveal preserved left ventricular wall motion with a dilated right heart associated with an elevated central venous pressure and low pulmonary artery pressures (Table 5.6). The failing right heart is unable to move blood through the pulmonary vasculature to the functional native left heart.

Medical therapy for right heart failure includes isoprotenerol hydrochloride (0.01-0.20 μg/kg/min IV), dobutamine hydrochloride (2-20 μg/kg/min IV) or milrinone lactate (50 μg/kg IV loading dose, 0.375-0.75 μg/kg/min IV continuous infusion maintenance dose). Intravenous nitroglycerin (0.1-30 μg/kg/min IV) or inhaled nitric oxide (20-60 ppm) selectively lower the pulmonary vascular resistance. The reduction in right ventricular afterload may increase forward flow through the right heart. The patient with medically refractory right heart failure becomes a candidate for a right VAD (RVAD). The patient can be supported with an Abiomed or Thoratec blood pump. Blood is withdrawn from the right atrium and returned to the proximal portion of the main pulmonary artery. As RVAD flow is initiated and CPB flow is reduced, systemic perfusion improves as blood moves across the pulmonary vasculature to the left heart.

The expectation is that right ventricular recovery will occur within 3-5 days. The time course of right ventricular recovery is accelerated when compared to the time course for left ventricular recovery. When compared to patients requiring left ventricular assistance, patients requiring right ventricular assistance alone are more easily weaned from mechanical circulatory support. If invasive hemodynamic monitoring is available, a pulsatile wave form is sought in the pressure trace from the pulmonary artery catheter as RVAD flow is reduced. If the native right ventricle is able to provide satisfactory left heart filling, pulmonary artery pressures and systemic perfusion will be maintained. If a right heart catheter is not in place persistent right ventricular dysfunction is manifest by the immediate appearance of jugulovenous distension as RVAD flow is reduced. When right ventricular recovery is complete the RVAD is removed. This is a relatively simple undertaking as CPB is not required.

An unusual form of "primary" right ventricular failure may occur in the patient with ischemic cardiomyopathy who is being supported with an LVAD as a bridge to cardiac transplantation.[7] If the patient has been stable for weeks or months on left heart support but suddenly develops reduced LVAD flow accompanied by chest pain and inferior wall ischemic changes on the electrocardiogram, treatment for right ventricular ischemia should be instituted. The acute decline in right ventricular function associated with a right ventricular infarction can reduce pulmonary blood flow and compromise LVAD filling. Coronary angiography confirms the diagnosis. Treatment protocols are similar to those provided for a patient who is not supported with an LVAD. If an LVAD is to be implanted in a patient with a critical right coronary artery stenosis, concomitant right coronary artery bypass grafting may protect the right ventricle during the period of LVAD support.

REFERENCES

1. Oz MC, Rose EA, Slater J et al. Malignant ventricular arrhythmias are well tolerated in patients receiving long-term left ventricular assist devices. J Am Coll Cardiol 1994; 24:1688-1691.

2. Kulick DM, Bolman RM III, Salerno CT et al. Management of recurrent ventricular tachycardia with ventricular assist device placement. Ann Thorac Surg 1998; 66:571-573.

3. Skinner JL, Bourge RC, Shepard RB et al. Simultaneous use of an implanted defibrillator and ventricular assist device. Ann Thorac Surg 1997; 64:1156-1158.

4. Hagley MT, Lopez-Candales A, Phillips KJ et al. Thrombosis of mitral valve bioprosthesis in patients requiring circulatory assistance. Ann Thorac Surg 1995; 60:1814-1816.

5. Dembitsky ZY, Buckley J, Jaski BE et al. The role of native valve mitral regurgitation during LVAD support. ASAIO J 1998; 44:17A.

6. Adamson RM, Dembitsky WP, Jaski BE et al. Left ventricular assist device support of medically unresponsive pulmonary hypertension and aortic insufficiency. ASAIO J 1997; 43:365-369.

7. McBride LR, Ruggiero R, Powers KA et al. Right ventricular infarction during left ventricular assist system support. Ann Thorac Surg 1998; 64:839-841.

7

Anesthesia for Patients with Ventricular Assist Devices

Javier H. Campos

INTRODUCTION

With the number of potential heart transplant recipients exceeding the number of donor hearts available, it is estimated that up to 30% of patients listed for cardiac transplantation die while waiting for a donor heart. In addition, 1-2% of the patients undergoing open heart surgery cannot be weaned from cardiopulmonary bypass (CPB). Mechanical ventricular assist devices (VADs) have been used to support the circulation in patients waiting for a donor heart and in patients recovering from CPB and cardiac surgery.[1] The blood pumps described elsewhere in this text can be used to support the left heart (left VAD, LVAD), right heart (right VAD, RVAD) or two pumps can be used to provide biventricular assistance (biventricular assist device, BVAD).

The clinical presentation of patients facing the cardiovascular anesthesiologist include the following:
- Deterioration of left ventricular function in a patient waiting for a donor heart transplant and requiring an LVAD.
- Patient undergoing open heart surgery, either first time or reoperation, with inability to wean from CPB during the operation.
- A patient with either an LVAD, RVAD or BVAD in place scheduled for heart transplantation.
- A patient scheduled to be weaned from the LVAD, RVAD or BVAD.

Mechanical Circulatory Support, edited by Wayne E. Richenbacher. © 1999 Landes Bioscience

ANESTHETIC MANAGEMENT

PREOPERATIVE EVALUATION

While waiting for a donor heart a potential cardiac transplant recipient may deteriorate hemodynamically. The patient's circulation is initially maintained with high dose inotropes (two or more) and possibly an intraaortic balloon pump (IABP). If the patient deteriorates further and requires VAD implantation, the patient will present for surgery with markedly altered hemodynamics typically characterized by a resting tachycardia and elevated central venous (> 20 mm Hg), pulmonary artery (pulmonary artery diastolic > 35 mm Hg) and pulmonary capillary wedge pressures consistent with a significantly diminished ejection fraction and low cardiac output.

Special attention is given to cardiovascular and pulmonary function. A concern for the anesthesiologist is the recognized defect of adrenergic mechanisms that modulate the heart's inotropic state. A reduction of beta$_1$ adrenergic receptor density in the failing heart has been demonstrated. This finding is thought to represent receptor down regulation produced by chronic exposure of the heart to excessive catecholamine levels. Although myocardial norepinephrine stores are decreased in advanced heart failure, circulating norepinephrine levels are elevated. The majority of patients scheduled for VAD(s) implantation are receiving high doses of inotropes. As these drugs act by stimulation of beta adrenergic receptors, maximal doses of inotropes often produce little effect on blood pressure and cardiac output. Usually more than two inotropes are given at any time and higher doses will be required during the administration of anesthetics.

Patients who are scheduled for VAD(s) implantation while waiting for a donor heart usually have a chronically dilated heart. They no longer effectively follow the Frank-Starling curve when subjected to additional volume loading. In fact, the failing ventricle is often regarded as insensitive to changes in preload and its stroke volume is described as fixed. Preload should be maintained during periods of venodilation induced by anesthetic agents. Patients with failing hearts usually have a very high afterload. Thus, increased chamber dimensions and chronic elevation of the aortic impedance spectrum combined with depressed contractility further impair the ventricle's basal performance and render it intolerant to additional afterload stresses that may be incurred during surgical stimulation. Any maneuver that raises systemic vascular resistance and systemic blood pressure without simultaneously augmenting myocardial contractility can produce a drastic fall in stroke volume, cardiac output and systemic perfusion.

PREMEDICATION

Patients scheduled for VAD implantation as a bridge to cardiac transplantation will usually be having elective surgery, therefore nothing per mouth after midnight is ordered. Generally, all cardiovascular medications that the patient is receiving on the inpatient unit should be given the morning of surgery; the only exception is that diuretics may be withheld. Premedication of these patients begins with sedatives (benzodiazepines). To decrease anxiety, sedatives are titrated upon insertion of intravenous lines and initiation of invasive monitoring in the operating room. Oxygen via nasal cannula or face mask is given to all patients. For patients that present to surgery with increased gastric contents, drugs that reduce gastric acidity or facilitate gastric emptying are used (metoclopramide and H_2 blocker such as ranitidine).

Example of premedication: 1) Supplemental oxygen
2) Midazolam hydrochloride 0.05 mg/kg titrated dose

MONITORING AND LINES PLACEMENT (TABLE 8.1)

1) Two large-bore peripheral intravenous lines are placed. One should be an 8.5 F rapid infuser catheter allowing rapid infusion of blood products in a short period of time. The second line is a 14 or 16 gauge catheter. These intravenous lines should be routed through a fluid warmer device.

2) Electrocardiographic monitoring with five leads (ECG) is used. Leads II and V_5 are monitored simultaneously intraoperatively. Computer analysis of ST changes are recorded.

3) A radial artery catheter is placed before induction of anesthesia. The majority of these patients have high afterload and peripheral vasoconstriction may be a problem. Occasionally, a femoral artery catheter placed under local anesthesia will be necessary. Another alternative is a brachial artery catheter.

4) A pulmonary artery catheter (Swan-Ganz), with or without oximetric capabilities, is placed via the right internal jugular vein if the patient is scheduled for VAD implantation alone. If the patient has a VAD in place and is to undergo a heart transplant, line placement is performed via the left internal jugular vein. The right internal jugular vein is preserved for subsequent endomyocardial biopsies. In addition, a double or triple lumen central venous pressure catheter may be

---

Table 8.1. Perioperative monitoring for patients requiring ventricular assist devices

1. Electrocardiogram (leads II and V_5)
2. Pulse oximetry (2 probes)
3. Arterial line (radial, brachial or femoral)
4. End tidal CO_2 monitor
5. Central venous pressure
6. Pulmonary artery catheter
    Measure pulmonary artery pressures (systolic, mean, diastolic, pulmonary capillary pressures)
    Systemic vascular resistance
    Pulmonary vascular resistance
    Stroke volume, left and right ventricular stroke work index
    Cardiac output (index)
    Mixed venous oxygen saturation
7. Left atrial pressure (intraoperative and postoperative)
8. Urine output
9. Transesophageal echocardiography

used for infusion of inotropes, vasopressors and vasodilators. For patients undergoing heart transplantation with a VAD in place, a long sterile sheath is used as the pulmonary artery catheter is withdrawn prior to excision of the native heart.

5) The transducer set-up must include four transducers as recordings of intracardiac pressures are performed by the cardiac surgeon when weaning the patient from CPB. The other three transducers are used for recording systemic arterial pressure, pulmonary artery pressure and central venous pressure.

6) A transesophageal echocardiogram (TEE) probe is placed after the patient's trachea has been intubated. A complete intraoperative TEE evaluation is performed prior to VAD insertion and following VAD implantation to assess device position and function. TEE is also important in the patient who is to undergo device explantation and cardiac transplantation.

7) Urinary drainage catheter to assess urine output is placed usually after the patient is anesthetized.

8) External defibrillator pads should be placed prior to induction of anesthesia in all patients who have had previous cardiac surgery.

INDUCTION AND MAINTENANCE OF ANESTHESIA

8

INDUCTION OF ANESTHESIA
Due to the severity of hemodynamic instability and the individual patient response, induction of anesthesia is often initiated with titrated doses of opioids. Fentanyl citrate (10-20 μg/kg IV) along with a hypnotic drug such as etomidate (0.1-0.3 mg/kg IV) can be used during induction. An alternative is ketamine hydrochloride. End organ unresponsiveness due to depletion of myocardial catecholamines and beta receptor down regulation can unmask ketamine hydrochloride's direct negative inotropic effect. If ketamine hydrochloride is chosen a dose of 0.5-1 mg/kg should be titrated intravenously. To facilitate tracheal intubation nondepolarizing muscle relaxants are used. Many of the patients who require VAD(s) implantation have impaired renal function and in these cases cisatracurium besylate is used. If renal function is preserved then rocuronium bromide (0.05 mg/kg) or vecuronium bromide is acceptable.

MAINTENANCE OF ANESTHESIA
There are only a few studies reporting the effects of intravenous agents (thiopental sodium and propofol) or inhalational agents (halothane and isoflurane) in humans with an artificial heart.[2-3] No studies describe the effect of these agents in patients supported with a VAD. Both thiopental sodium (5 mg/kg) and propofol (2.5 mg/kg) have been shown to have potent vasodilator effects on venous and arterial beds in patients with artificial hearts. Furthermore, the effects on halothane and isoflurane at minimal alveolar concentrations of 1 and 1.5% have shown profound arterial and venodilator properties in patients with artificial hearts.

Knowing the severity of the hemodynamic instability in the presence of intravenous or inhalational agents, it appears justified to use very low concentrations of inhalation agents such as isoflurane at concentrations of 0.5%. These low concentrations should prevent awareness during anesthesia. Another alternative is to use scopolamine hydrobromide (0.2-0.4 mg IV) in conjunction with a titrated dose of narcotics (fentanyl citrate) during maintenance of anesthesia or titrated doses of benzodiazepines such as midazolam hydrochloride.

Ventilatory management during induction of anesthesia is also important. Once airway management is established the peak airway pressures should not exceed 25 cm H_2O as high intrathoracic pressures may decrease venous return in patients who already have some form of right ventricular dysfunction.

PLACEMENT OF LEFT VENTRICULAR ASSIST DEVICE AND ANESTHETIC IMPLICATIONS

The various devices used as LVADs have already been explained in great detail elsewhere in this text.[4-6] In general, these devices are placed with the patient on CPB. Preoperative assessment with TEE is used to assess valve lesions. Aortic insufficiency must be addressed during LVAD insertion. If the patient has aortic insufficiency some of the blood pumped into the aortic root by the LVAD will flow backward across the aortic valve decreasing net forward flow. If severe aortic insufficiency is present vital organ perfusion can be compromised. Mitral insufficiency is not an important consideration in LVAD recipients. As right ventricular failure is the cause of death in 20% of LVAD recipients, every effort should be made to correct tricuspid insufficiency prior to LVAD placement. A complete assessment of cannulae insertion sites (inflow and outflow) should be made with TEE. Atheromatous plaques in the ascending aorta and ventricular thrombus should be identified. Avoidance of such abnormalities during device insertion reduces the potential for subsequent thromboembolic events. Weaning a patient with an LVAD from CPB requires deairing of the VAD and the heart as well as inflow and outflow cannulae. The best method to confirm that air is no longer present is an intraoperative TEE. The head down position along with Valsalva maneuvers facilitates deairing the heart and LVAD.

Patients supported with an LVAD may require inotropes prior to coming off CPB (Table 8.2). Systemic vasodilation may be a problem in this situation if the left atrial pressure is normal and systemic blood pressure low. Administration of moderate doses of norepinephrine bitartrate (0.05-0.1 µg/kg IV) may be necessary and on occasion more than one inotrope may be needed. If right ventricular dysfunction is present, but not severe enough to require an RVAD, administration of amrinone lactate or dobutamine hydrochloride may improve right ventricular function. Bleeding and coagulopathy is a concern after LVAD placement. Packed red blood cells (PRBC), fresh frozen plasma (FFP) and platelets are often necessary. Hemofiltration may be used after the patient has been weaned from CPB. Hemofiltration decreases third space fluid accumulation and reduces the potential

Table 8.2. Drugs needed for ventricular assist device placement

Drug	Preparation	Concentration	Loading dose	Infusion
Inotropes				
Epinephrine	2 mg diluted in 250 ml NS	8 µg/ml		0.05-0.2 µg/kg/min
Norepinephrine bitartrate	2 mg diluted in 250 ml NS	8 µg/ml		0.01-0.1 µg/kg/min
Isoproterenol hydrochloride	2 mg diluted in 250 ml NS	8 µg/ml		0.01-0.05 µg/kg/min
Dopamine hydrochloride	200 mg diluted in 250 ml NS	800 µg/ml		3-10 µg/kg/min
Dobutamine hydrochloride	250 mg diluted in 250 ml NS	1000 µg/ml		5-10 µg/kg/min
Amrinone lactate	250 mg diluted in 250 ml NS	1000 µg/ml	1 mg/kg (undiluted)	5-10 µg/kg/min
Milrinone lactate	25 mg diluted in 250 ml NS	100 µg/ml	50 µg/kg over 10 min	0.35-0.75 µg/kg/min
Vasodilators				
Nitroglycerin	50 mg diluted in 250 ml NS	200 µg/ml		0.5-8 µg/kg/min
Sodium nitroprusside	50 mg diluted in 250 ml NS	200 µg/ml		0.5-8 µg/kg/min
Vasopressors				
Phenylephrine hydrochloride	20 mg diluted in 250 ml NS	80 µg/ml	1-5 µg/kg	
Pulmonary vasodilator				
Nitric oxide	gas mixture (respiratory therapy)	inhaled 20-40 ppm		

NS = normal saline

8

for pulmonary and other complications related to interstitial edema. Reversal of heparin sodium is performed once hemodynamic stability is achieved after CPB has been discontinued. A test dose of protamine sulfate is administered and if no reaction occurs the full dose of protamine sulfate is slowly given.

PLACEMENT OF RIGHT VENTRICULAR ASSIST DEVICE AND ANESTHETIC IMPLICATIONS

It is estimated that up to 20% of patients supported with an LVAD may develop severe right ventricular dysfunction and require RVAD implantation. Cannulation for right heart support uses right atrial cannulation with return to the pulmonary artery. Evaluation of the cannulation sites with TEE is mandatory. Inspection for pre-existing thrombus along with tricuspid valve assessment and visualization of the pulmonary artery trunk is made. Patients who require an RVAD usually need inotropes that work better on the right side of the heart such as dobutamine hydrochloride or amrinone lactate. Vasodilators such as nitroglycerin or prostaglandins are administered to patients who have some degree of pulmonary hypertension. Nitric oxide administration may be necessary while the patient is being weaned from CPB and thereafter. In our center, nitric oxide is delivered via a system that is attached to the respiratory limb of the anesthesia breathing circuit. The recommended initial dose is 10-20 ppm. Constant evaluation of serum methemoglobin levels is necessary while nitric oxide is being administered.

PLACEMENT OF THE BIVENTRICULAR ASSIST DEVICES AND ANESTHETIC IMPLICATIONS

BVADs are used for patients with left and right ventricular dysfunction who are unable to be weaned from CPB or those with progressive postcardiotomy cardiac failure in the postoperative period. The extracorporeal VADs can be configured for both right and left heart support. For left ventricular assistance, the inflow cannula is inserted via the left atrium while the outflow cannula is placed in the ascending aorta. For the right side of the heart the cannulation sites are the right atrium for the inflow cannula and the pulmonary artery for the outflow cannula. Patients who are undergoing BVAD placement are on maximal doses of inotropes and vasodilators. These patients quite often will be receiving nitric oxide as right ventricular dysfunction with pulmonary hypertension is so severe. The survival rate for patients requiring BVAD is lower than for patients requiring univentricular support (RVAD or LVAD). The decreased survival with BVAD is related to the degree and extent of ventricular damage. Biventricular support for postcardiotomy cardiogenic shock is associated with a lower survival rate (19.7%)

than that seen in patients requiring isolated left (27.7%) or right (25.6%) ventricular assistance.

APROTININ IN PATIENTS REQUIRING A VENTRICULAR ASSIST DEVICE

Bleeding and coagulation disorders are a common problem in patients undergoing VAD placement, especially those who have had a prior cardiac surgery. Aprotinin, a bovine serine protease inhibitor, has been used extensively in patients undergoing cardiac operations in an effort to minimize perioperative blood loss, prevent potential complications associated with blood transfusions and the need for reoperations for control of bleeding. Aprotinin inhibits the activity of kallikrein and plasmin, thereby effectively serving as an antifibrinolytic agent.

In a retrospective study, aprotinin was used successfully in patients undergoing LVAD placement.[7] Postoperative bleeding assessed by chest tube output was 2.5 times less in those patients receiving aprotinin perioperatively compared to a control group. In addition, the number of units of PRBC, FFP, cryoprecipitate and cell saver blood required intraoperatively was significantly reduced.

Points of concern with the use of aprotinin include:

1) Hypersensitivity and anaphylactic reactions are rare ($< 0.1\%$) in patients with no prior exposure to aprotinin. However patients previously exposed to aprotinin are at increased risk for developing severe anaphylactic reactions. Attention must be given to the fact that a patient with previous exposure to aprotinin can present a severe reaction if it is used again during LVAD placement or removal.

2) Another potential problem associated with the use of aprotinin is the appearance of renal dysfunction that can appear days later in the postoperative period. A study had shown increases in serum creatinine levels (50% elevation over baseline) in 39.5% of the patients that received aprotinin.[8] However, none of the patients required permanent hemodialysis.

Aprotinin administration in VAD patients is summarized in Table 5.3. Patients requiring a VAD have a high risk of bleeding and bleeding-related morbidity/mortality. VAD placement then further compounds the pre-existing hemostatic derangement with abnormal flow conditions introduced by the device and by biomaterial surfaces that interact with blood. It is important to achieve hemostasis rapidly with early and vigorous infusions of platelet concentrates and FFP before prolonged bleeding results in uncontrollable coagulopathy.

For an elective VAD placement the blood bank should be alerted to initially order 10 units of PRBC, 4 units of FFP and two 8 packs of platelets. In addition, 10 units of cryoprecipitate may be necessary. If blood products are required, serial coagulation tests are performed including prothrombin time (PT), partial thromboplastin time (PTT), platelet count and fibrinogen studies.

INTRAOPERATIVE TRANSESOPHAGEAL ECHOCARDIOGRAPHIC EVALUATION IN PATIENTS WITH VENTRICULAR ASSIST DEVICES

Prior to the initiation of CPB a baseline TEE study is performed.[9,10] The TEE probe is passed after induction of anesthesia and the patient's trachea has been intubated. The first evaluation consists of a transverse, four chamber view (long axis view). This view allows evaluation of left and right ventricular filling and detects the presence of ventricular aneurysms, ventricular or atrial thrombus and pericardial effusions. Right and left atrial size is assessed and the atrial septum is evaluated with color flow doppler for the presence of a patent foramen ovale (PFO) or atrial septal defect (ASD). The ventricular septum is also evaluated with color flow doppler for the presence of ventricular septal defects. Presence of a PFO is a well known cause of hypoxemia in patients with an LVAD.[11] One of the hemodynamic effects of an LVAD is to reduce the left atrial pressure. If the right atrial pressure exceeds the left atrial pressure a right-to-left shunt occurs. A contrast test to inspect for a PFO should be performed in every patient that requires a VAD. Agitated normal saline, 8 ml mixed with 2 ml of the patient's blood, is injected into the right atrium. A contrast image is obtained looking for microbubble transport to the left side through a PFO or ASD.

The next evaluation is the basal short axis view to visualize the base of the heart and the great vessels. This image plane is optimal for viewing the aortic valve cusps. The aortic valve leaflets should be inspected. Color flow doppler can be utilized in this view as a preliminary screen for aortic insufficiency. Imaging of the aortic valve, left ventricular outflow tract, ascending thoracic aorta and right sided cardiac structures are obtained at this level by rotating the adjustable scan angulation feature of the multiplane probe from 0°-130°. Attention should be given to the presence of atheromatous plaques in the ascending thoracic aorta or the presence of thrombi in any view of the heart.

The basal short axis view angulating the probe from 0-130° can provide different views.

- Bicaval view: superior and inferior vena cavae, left and right atrium, interatrial septum, right pulmonary artery and right pulmonary vein.
- Ascending thoracic aortic view: ascending thoracic aorta, left and right ventricular outflow tracts.
- Wrap around view: right-sided cardiac structures (right atrium, right ventricle, tricuspid valve, right ventricular outflow tract, pulmonic valve and main pulmonary artery). All these structures wrap around a cross section view of the aortic valve.
- Mitral valve view: anterior and posterior mitral valve leaflets, left atrium and left atrial appendage, left ventricle and left ventricular outflow tract along with left pulmonary veins.

The transgastric short axis view of the ventricles inspects for size, wall thickness, overall contractility and regional wall motion abnormalities. A severely dilated left ventricle with an akinetic apex may harbor a mural thrombus. One of the most important features to be evaluated preimplantation is right ventricular function and potential need for mechanical right ventricular support. The last observation is of the ascending aorta, aortic arch, descending thoracic aorta and evaluation for atheromatous plaque formation, thrombus or the presence of an aortic aneurysm.

For postimplantation assessment of the LVAD with TEE a transverse four chamber view is very helpful in showing the correct positioning of the apical conduit (Figs. 8.1, 8.2). Color flow doppler and continuous wave doppler must demonstrate a laminar flow pattern into the cannula. The response of the right ventricle to the implantation of the LVAD is one of the key features determining the overall success of the implantation procedure. A complete assessment of right ventricular function is evaluated. The tricuspid valve is assessed with color flow doppler. LVAD insertion increases right ventricular preload and improves right ventricular compliance, therefore, right ventricular end-diastolic volume increases while right ventricular end-diastolic pressure decreases. As a result, LVAD implantation may acutely decrease mitral regurgitation while worsening tricuspid regurgitation. This is probably secondary to a leftward shift of the interventricular septum, increased right ventricular compliance and an increasing volume load to the right ventricle with increased cardiac output. Right atrium and ventricle size, along with right ventricular wall thickness and contractility, are evaluated by measuring right atrial diastolic function including evaluation of hepatic veins and total flow velocities. A view of the atrial septum in conjunction with color flow imaging should be able to detect any PFO if present, once implantation of the LVAD has occurred. Again an agitated normal saline contrast test should be performed (Fig. 8.3).

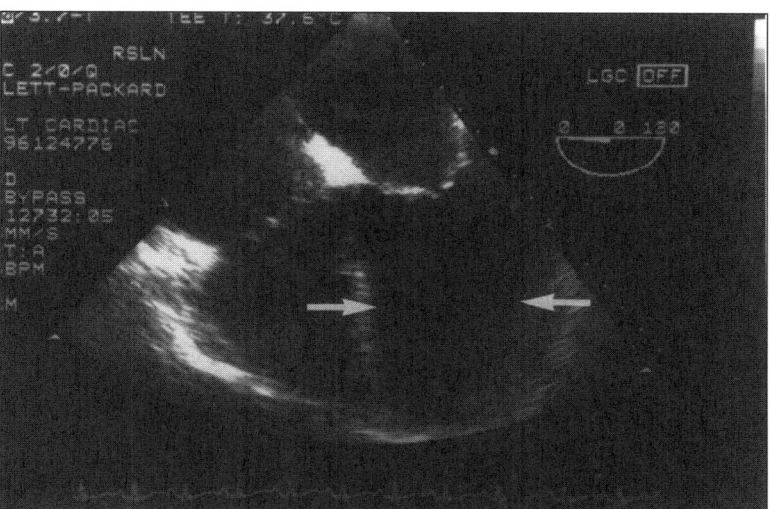

Fig. 8.1. Intraoperative transesophageal echocardiographic images showing midesophageal four chamber view. The arrows indicate a dilated left ventricle before implantation of a left ventricular assist device in a patient with dilated cardiomyopathy.

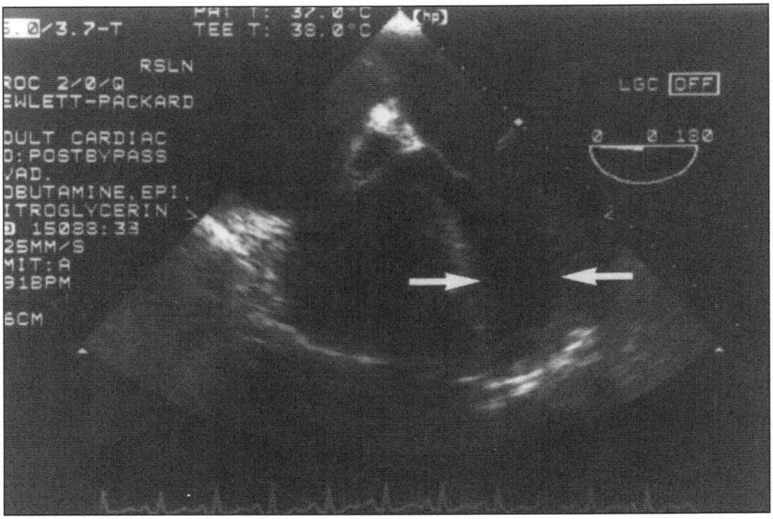

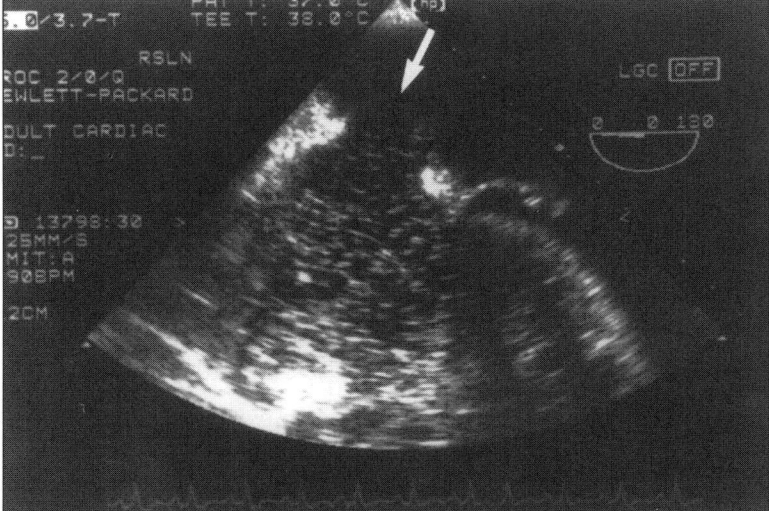

Fig. 8.2. Midesophageal four chamber view showing a decompressed left ventricle after implantation of a left ventricular assist device.

Fig. 8.3. Intraoperative transesophageal echocardiographic image showing injection of agitated normal saline. Microbubbles are seen in the right atrium. The arrow shows the atrial septum. This test should be performed after implantation of a left ventricular assist device to detect a patent foramen ovale.

TRANSPORT OF PATIENTS WITH VENTRICULAR ASSIST DEVICE TO THE SURGICAL INTENSIVE CARE UNIT

Patients who undergo VAD placement remain intubated in the immediate postoperative period. Transport from the operating room to the surgical intensive care unit is made taking all precautionary measures for unexpected or unwanted complications during transport. An airway management kit should include laryngoscope blade, handle, mask and endotracheal tubes. A transport monitor must include assessment of the ECG, systemic arterial pressure, pulmonary artery pressure or left atrial pressure and pulse oximetry. A pacemaker pulse generator and resuscitation drugs including epinephrine, calcium chloride and sodium bicarbonate should be included in the transport. VAD patients may continue to bleed in the early postoperative period, therefore, blood products should be carried during transport. Close communication between the anesthesiologist, cardiac surgeon and perfusionist should exist so that all are aware of the VAD triggering mode used during transport. If a console has to be moved while transporting the patient, coordination must be maintained to avoid mishaps, like loosening inflow or outflow cannulae. Ventilation should be gentle to avoid an exacerbation of right ventricular dysfunction. Multiple vasoactive drug infusions are usually given to the patient. Therefore, extra personnel are needed to help with the device(s) or to push infusion pumps or an IV pole. As ongoing bleeding may occur in the VAD patient, an anesthesia machine and drugs should be ready in the operating room in the event mediastinal exploration is necessary.

REFERENCES

1. Ott RA, Mills TC, Eugene J. Current concepts in the use of ventricular assist devices. Cardiac Surgery: State of the Art Reviews 1989; 3:521-542.
2. Rouby J-J, Andreev A, Léger P et al. Peripheral vascular effects of thiopental and propofol in humans with artificial hearts. Anesthesiology 1991; 75:32-42.
3. Rouby J-J, Léger P, Andreev A et al. Peripheral vascular effects of halothane and isoflurane in humans with an artificial heart. Anesthesiology 1990; 72:462-469.
4. Oz MC, Rose EA, Levin HR. Selection criteria for placement of left ventricular assist devices. Am Heart J 1995; 129:173-177.
5. McCarthy PM, James KB, Savage RM et al. Implantable left ventricular assist device. Approaching an alternative for end-stage heart failure. Circulation 1994; 90 [part II]: II-83-II-86.
6. Burton NA, Lefrak EA, Macmanus Q et al. A reliable bridge to cardiac transplantation: The TCI left ventricular assist device. Ann Thorac Surg 1993; 55:1425-1431.
7. Goldstein DJ, Seldomridge JA, Chen JM et al. Use of aprotinin in LVAD recipients reduces blood loss, blood use, and perioperative mortality. Ann Thorac Surg 1995; 59:1063-1068.
8. Goldstein DJ, Oz MC, Smith CR et al. Safety of repeat aprotinin administration for LVAD recipients undergoing cardiac transplantation. Ann Thorac Surg 1996; 61:692-695.

9. Savage RM, McCarthy PM, Stewart WJ et al. Intraoperative transesophageal echocardiographic evaluation of the implantable left ventricular assist device. Video J Echo 1992; 2:125-136.

10. Faggian G, Dan M, Bortolotti U et al. Implantation of an external biventricular assist device: Role of transesophageal two-dimensional echocardiography. J Heart Transplant 1990; 9:441-443.

11. Baldwin RT, Duncan JM, Frazier OH et al. Patent foramen ovale: A cause of hypoxemia in patients on left ventricular support. Ann Thorac Surg 1991; 52:865-867.

8

Nursing Care of the Patient Requiring Mechanical Circulatory Support

Kelly L. Jones, Carolyn J. Laxson, Sarah C. Seemuth, Sara J. Vance

NURSING STAFF EDUCATION

The first step in initiating a mechanical circulatory support program is to develop a multidisciplinary team. An important part of the team is the nursing staff comprised of individuals from a variety of patient care areas. At the outset it must be determined if the entire unit staff from each care area is to be trained or if a dedicated core group of nurses will be selected. This decision is based on the following factors: staff seniority, expected vacancies, open positions, patient census, patient diversity and the expected volume of ventricular assist device (VAD) patients. If a unit chooses to begin with a core group, coverage must be guaranteed on weekends, off shifts and during vacations. Once oriented the core group acts as preceptors for untrained or new staff members.

The initial orientation to the use of a VAD system is provided by the device manufacturer. Most companies employ clinical application specialists who provide inservicing and clinical support to hospitals using their device. Occasionally, a company employs a nurse from another hospital who has advanced training and knowledge of their device to provide orientation at new centers. These inservices are helpful and necessary but are **not** a substitute for additional, more specific training that should follow. Topics covered during a manufacturer's inservice include the history and development of the device, application for use, clinical trial results, study publications, patient selection criteria, potential complications, patient care issues and instructions regarding the equipment. As a great deal of information is presented during the overview inservice, there is often insufficient time for staff to obtain the hands-on training needed to be competent with the equipment. One way to address this problem is to have the company representative provide separate inservices for the operating room, intensive care unit and step-down unit staff. Focusing the discussion on issues pertinent to a specific patient care area enhances the educational experience and allows the nursing staff to spend more hands-on time with the VAD equipment.

Subsequent inservicing is provided by nursing staff who have often attended an off-site training program provided by the device manufacturer. The individual responsible for orienting the nursing staff varies from program to program. This task is assigned to a senior staff nurse, assistant nurse manager, transplant coordinator or a VAD coordinator. The decision depends upon the size of the program and personnel available. When planning orientation consider having individual(s) from each unit orient their own personnel (operating room, intensive care unit, step-down unit). This is more efficient than using a single person as it provides staff with more accessibility to the instructor and a more unit-specific orientation. Subsequent inservicing should not mirror the initial clinical application specialist's presentation. Rather, it focuses more on direct patient care, equipment and nursing responsibilities. Developing a simple condensed outline for staff to follow is key to meeting the goals of the inservice. Although all staff should have ready access to the VAD users manual and additional in-depth resources concerning VAD use, it is not necessary, and in fact counterproductive, to provide everyone with a copy of the users manual for training purposes. Topics discussed during the inservice sessions include those pertinent to any patient supported with a VAD regardless of the manufacturer, and device specific considerations with emphasis on device function and troubleshooting (Table 9.1).

Begin with a review of circulatory physiology in VAD patients. A generic review of VAD design and function provides the nursing staff with a better understanding of how the blood pump interacts with the patient's heart. An example illustrating this point is the hemodynamic response to hypovolemia in a patient

9

Table 9.1. Inservice essentials

Blood pump physiology
    Device design and function
    Atrial versus ventricular cannulation for left ventricular assistance
    Interaction of the device with the heart
        Single versus biventricular failure
        Arrhythmias
Patient care
    Indications for ventricular assistance
    Patient selection criteria
    Perioperative management
    Long-term care issues
    Drive line/cannulae exit site care
    Documentation in the medical record
Device operation
    Power source
    Modes of operation
    Console operation
    Hands-on operation
Troubleshooting
    Alarm responses
    Back-up procedures
    Patient care scenarios

supported with a VAD versus the response seen in a patient with normal physiology. With hypovolemia the heart rate increases to maintain systemic perfusion. In the patient supported with a VAD, however, hypovolemia is manifest by a decrease in VAD pumping rate. As the VAD pumps a full stroke volume in the automatic mode, the pumping rate slows to allow time for the VAD to fill in the hypovolemic patient.

The patient management discussion is tailored to suit the needs of the particular patient care area. Routine nursing assessments require modification. For example, heart and lung sounds are very difficult, if not impossible, to hear due to competing sounds from the pump. Patients supported with a mechanical blood pump are tolerant of the most malignant of arrhythmias. "No chest compressions" is a routine order. Drive line/cannulae exit site care instructions are thoroughly reviewed. The selection of problems for in-depth discussion is tailored to each unit's inservice depending upon the likelihood of occurrence. For example, the risk of developing right heart failure or severe bleeding is higher in the early postoperative stages while the patient is in the operating room or the intensive care unit and is discussed more thoroughly with these respective units.

Instructions regarding device operation comprises the majority of the inservice. Duplicate the actual equipment used whenever possible. In some cases a mock circulatory loop may be available to simulate the entire system. If VAD system components cannot be used during the inservice, a graphic displaying the entire system with the individual parts identified is helpful. Each piece of equipment and its function are demonstrated and discussed. Simulation and hands-on exposure combined with practice is the best method for teaching new skills.[1] It is important to understand normal VAD operation before instruction can proceed to alarm response and system troubleshooting. Each VAD system has a means for monitoring its own operation. Abnormal device function generates an alarm condition. Depending upon VAD type there may be one universal alarm regardless of the problem or there may be different alarms for each type of problem or degree of urgency of the problem. Check with the device manufacturer to see if patient education material includes an alarm algorithm. Such an algorithm helps staff responding to an alarm condition act in a timely and systematic manner. If such an algorithm is not available one should be developed keeping directions simple and concise. An example of an alarm algorithm developed for use with the Thermo Cardiosystems VE LVAS (Thermo Cardiosystems, Inc., Woburn, MA) is attached (Table 9.2). Copies of the algorithm should be kept in the patient's room and with travel supplies for use when the patient leaves the primary care area.

When a problem causes an alarm to be activated immediate attention is necessary. Ideally, two staff will respond to an alarm situation. One staff provides instructions using the algorithm while the second staff carries out the instructions. The first response in any alarm algorithm is to evaluate the patient and pump. If the pump has stopped emergency interventions are initiated. Each VAD system can be powered by manual equipment in the event emergency pump operation is necessary. The manual equipment varies with device manufacturer and includes a hand-pump, foot-pump or cranking handle all of which generate pneumatic

---

Table 9.2. Alarm algorithm for Thermo Cardiosystems VE LVAS

Initial response to all alarms:
1.   Obtain assistance from another nurse.
2.   Check vent line for obstruction.
3.   Check connection between patient and controller.
4.   Check connection between controller and power source.

If alarm stops no further action is necessary.

If alarm continues:

     Yellow wrench alarm
        1.   Assess patient's tolerance.
        2.   Assess device rate and flow if currently on monitor.
        3.   Change power source.
        4.   Connect to system monitor.
        5.   If possible change the controller.
        6.   Change power base unit cable.
        7.   Change power base unit.
        8.   Obtain additional help.
        9.   Consider attaching to pneumatic console.

     Red Heart Alarm (functioning pump and patient is stable)
        1.   Assess patient's tolerance.
        2.   Assess device rate and flow if currently on monitor.
        3.   Be prepared to hand-pump.
        4.   Change power source.
        5.   Change controller.
        6.   Obtain additional help.
        7.   Consider attaching to pneumatic console.

     Red Heart Alarm (pump is not functioning or patient is unstable)
        1.   Disconnect power source.
        2.   Hand-pump patient.
        3.   Obtain additional help.
        4.   Change power source.
        5.   Change controller.
        6.   Attach to pneumatic console.

---

pressure to initiate pump systole. Staff should be comfortable using and proficient with the emergency equipment. Having actual equipment available at the inservice for staff to learn and practice with is invaluable.

After completing the inservice staff competency is evaluated. To be certified the nurse must demonstrate sufficient knowledge of the VAD as defined by the objectives in Table 9.3. A written examination is used to determine comprehension of the material but it does not assess performance with the equipment. The use of actual equipment to demonstrate technical skills in created scenarios is a mandatory component of competencies. These methods are more time consuming to conduct than written tests. However, they offer a better assessment of nurses ability to apply the information. The purpose of this form of evaluation is to predict how the nurse will perform in the future.[1]

Competency is assessed periodically. The frequency with which competency is assessed varies with the program but should be accomplished at least annually.

Centers that perform only a few VAD implantations each year may find their staff having difficulty mastering and retaining skills due to limited exposure. In this situation competencies are needed more frequently to ensure staff remain qualified. Offering a review to the staff is desirable before competencies are given. The amount of time and information required is determined by the staff's experience. The need for additional review and training is also considered when there has been significant staff turnover. In addition to regular reviews and competencies it is important to offer the staff a variety of reference materials including the users manual, inservice outlines, algorithms and journal articles. Each center should develop a VAD protocol book that contains information relevant to the operation of the program within the institution (Table 9.4). The content of the VAD book may vary to reflect center specific practices.

Table 9.3. Competency objectives for the ventricular assist device

1. Explain pump circulation physiology and apply it to responses seen during patient conditions such as hypertension, hypovolemia, right heart failure and arrhythmias.
2. Identify the equipment and discuss/demonstrate its use and maintenance with normal operation.
3. Define different alarm conditions and discuss possible causes and corrections by utilizing the troubleshooting algorithm.
4. Recognize when to initiate emergency back-up measures and demonstrate use of back-up equipment.
5. Name additional resources to contact for help with troubleshooting or emergencies.
6. Demonstrate drive line/cannulae exit site dressing change procedure.

Table 9.4. Contents of ventricular assist device protocol book

Patient selection criteria
Preoperative evaluation tests
Perioperative antibiotic regimen
Standardized orders
     Preoperative
     Postoperative
     Transfer to general ward
Care map
Patient care procedures
     Drive line/cannulae skin exit site care
Patient education materials
Staff competency objectives
Policies
     Transportation
     Defibrillation/cardioversion
Community education materials
Outpatient follow-up
Contact numbers for VAD personnel

PREIMPLANTATION PSYCHOSOCIAL ASSESSMENT

When a patient's candidacy for mechanical circulatory support is assessed psychosocial factors are considered along with medical eligibility. A psychosocial assessment of the patient and family is an intervention used to gather information and establish a relationship with them. The family is defined as any person(s) who provides support to the patient. Information for the assessment is gathered from the patient, family, medical chart and previous service providers. A psychosocial assessment reviews many factors of the patient's life including current living situation, adjustment to illness, financial considerations, legal issues and mental status. An understanding of these factors assists the patient, family and healthcare team in identifying strengths and weaknesses of the patient. It will also assist in determining what further interventions are needed to facilitate the patient's candidacy for and success as a VAD recipient.

LIVING SITUATION

The practitioner should begin the assessment by gathering information about the patient and his family. The family genogram is a tool that provides a visual picture of the family and the patient's place within the family structure. The genogram involves three levels including mapping the family structure, recording family information and delineating family relationships. The genogram is one part of the clinical assessment and information from it should be integrated with the total family assessment. Along with understanding the structure of the family it is important to assess the patient's support system. Being a successful VAD patient requires an adequate support system provided by family, friends and/or organizations. These support resources must have a clear understanding of their role, tasks and their ability to perform them. A plan is developed outlining who will be doing what tasks and a verbal commitment is obtained from the responsible individual(s). The patient and family may need assistance from the healthcare team to develop a plan and follow it.

The physical structure of the patient's dwelling is discussed. Knowing the patient's accessibility to the bedroom, bathroom, living area, location and number of stairs and who resides with the patient helps identify discharge planning needs. Another aspect of discharge planning is the patient's ability to access the implant center. Transportation to the treatment center is critical for the VAD patient both for routine follow-up and acute care needs. The patient should have access to a reliable mode of transportation. At the least, a primary and back-up driver should be identified. The transportation plan includes access to the implant center within a time frame to be determined by the center.

Some patients need to relocate closer to the implant center if they live outside of the acceptable travel time. If relocation is required this is discussed with the patient and family. It may be helpful to network the patient and family with other patients who need to relocate. Relocation may be temporary or permanent depending upon the center's established guidelines.

ADJUSTMENT TO ILLNESS

The ability to adequately cope with a chronic illness and responsibly follow a medical treatment plan are requisite skills. How the patient and family believe the patient has coped with his illness in the past or coped with other difficult situations provides insight to how the patient may cope in the current situation. Obtaining a list of what has helped the patient cope in the past is useful in working with the patient now and in the future. Similar to coping, the patient's ability to follow his medical treatment plan in the past provides insight into future compliance. In what areas does the patient excel? What areas need improvement? Specific areas to explore include attending appointments, obtaining laboratory work, taking medications and asking questions of the healthcare team.

The VAD recipient and family are asked to identify their goals and expectations after VAD placement. This discussion allows the healthcare provider to correct any unrealistic beliefs, further educate the patient and family and describe what life may be like after VAD placement.

MENTAL STATUS

To further enhance the healthcare team's understanding of the patient's current mental health, a mental health history is obtained. This history includes any current concerns, diagnoses, previous treatments or use of psychotropic medications. Records pertaining to current or past treatment should be reviewed. The patient's exposure to alcohol, illicit drugs and tobacco is determined. This review includes current usage, past usage, any previous treatment and treatment outcomes. If there are concerns about the patient's mental health or history of substance abuse, consultation is obtained with a psychiatrist or chemical dependency counselor.

LEGAL

Patients who have had difficulties with chemical dependency may also have legal issues such as probation or pending criminal charges. The patient and family should be questioned regarding any probation or pending charges. If there are active legal issues further investigation must occur to determine what impact these would have on the patient's accessibility to VAD follow-up and care.

Patients should be provided with information about legal documents pertinent to their health. In the interview, information regarding advanced directives is reviewed with the patient and family. If available, a brochure or information sheet concerning Durable Power of Attorney and Living Will is provided. The patient is instructed about where to obtain forms and notary services in order to complete these documents.

FINANCIAL

Finances are a primary concern for the patient and his family. Payor sources for treatment including inpatient and outpatient care, supplies, medications and cardiac rehabilitation are investigated. If needed, precertification and preauthorization are obtained. Other alternative payors such as entitlement and categorical programs are discussed if applicable. Information regarding disability is provided if appropriate. The patient and family are encouraged to familiarize themselves with how to obtain benefit information, be aware of the referral or precertification processes, know his benefits and preferred provider requirements if applicable.

POSTOPERATIVE CARE

The postoperative care of a patient who has received a VAD is complex and requires the utmost use of nursing assessment and intervention skills (Table 9.5). Meticulous care must be paid to even the smallest detail in order to ensure an optimal patient outcome. Whether the goal is to support the patient pending a recovery of ventricular function or long-term support of the systemic circulation as a bridge to cardiac transplantation, excellent postoperative nursing care is key to assuring that goals are met.

Table 9.5. Nursing interventions for the patient requiring mechanical ventricular assistance

---

Blood pressure, pulse, respirations and temperature q4h while awake.
      Call physician for: Systolic blood pressure > _____ , < ____
                      Diastolic blood pressure > _____ , < ____
                      Temperature > _____
Ventricular assist device (VAD) parameters q4h while awake.
      Rate, volume, flow/output
      Call physician for VAD flow/output < __
Continuous cardiac monitoring. Discontinue on postoperative day _____ if stable.
Defibrillation/cardioversion precautions (electrically isolate device if necessary).
Intake/output every _____ h.
Weight QD.
Edema check every _____ h.
Skin/integument check every _____ h.
Sternal/abdominal incision site check q8h.
Site check—drains every _____ h.
Site check—peripheral intravenous catheter(s) every _____ h.
Drain-suction _____ mm Hg.
Drive line/cannulae/drains dressing changes every 8h.
      Observe patient/significant other until independent.
Emergency procedure readiness
      Manual pump, back-up equipment with patient at all times.
      Post pager numbers in room (perfusionist, surgeon, cardiologist, VAD coordinator).
Device routine daily/weekly maintenance checks (specify).
Activity level—up ad lib and secure drive line/cannulae when ambulating.
Cardiac rehabilitation QD.
Diet (specify diet, fluid restrictions, nutritional supplements, calorie counts).
Bath—independently QD, no showers.
Oral care—independently BID.

Pulmonary care
    Incentive spirometry every _____ h while awake.
    Cough and deep breathing every _____ h while awake.
    Chest percussion gentle to back **only** every _____ while awake.
    Continuous pulse oximetry/spot check oxygen saturation every _____
     while awake.
    Call physician for oxygen saturation < _____
    Breath sounds monitoring every _____ h while awake.
Pain monitoring q2h while awake and following intervention.
Laboratory work (specify) every _____ ; all labs in pediatric tubes.
Call light within reach, constant.
Instruction
    Diet QD and PRN.
    Procedures/treatments QD and PRN.
    Medications with every administration and PRN.
    Activity QD and PRN.
    With every dressing change PRN.
    Emergency VAD procedures and review QD.
Referrals
    Cardiac rehabilitation
    Dietitian
    Social worker
    Chaplain
    Other
Discharge planning activities QD
    Return demonstrate/verbalize VAD routine care.
    Return demonstrate dressing change.
    Return demonstrate emergency procedures (manual pumping and back-up
    equipment).
    Return demonstrate/verbalize/identify VAD equipment and function.
    Order dressing supplies prior to discharge.
    Order pager prior to discharge.
    Return demonstrate/verbalize how to contact VAD coordinator.
    Demonstrate completion of daily diary documentation.
     Vital signs, VAD flow, rate and stroke volume.
    Demonstrate correct self medication administration.

Pulmonary Care

Pulmonary care is vital to the patient's well-being as he is recovering from a sternotomy and with an implantable VAD, a midline abdominal incision as well. Provided the patient is hemodynamically stable extubation is usually possible within 12-18 hours after VAD insertion. Following extubation supplemental oxygen is administered via mask or nasal cannula. With diuresis the supplemental oxygen can be discontinued within a few days. Patients perform incentive spirometry and associated coughing and deep breathing hourly while awake. Assessment of lung sounds is difficult due to noise from the VAD so close observation of respiratory effort, color and oxygen saturation are essential. Chest physiotherapy is performed **gently** to the patient's back **only** if needed to assist in loosening secretions. Chest roentgenograms are initially performed daily to assess for atelectasis, infiltrate, effusion or pneumothorax. Mediastinal tubes are usually removed within the first 24 hours. Should mediastinal or pleural tubes be needed for a longer period of time, interval assessment is made for air leaks, amount and characteristics of drainage. In addition, adequate analgesics must be provided to enable patients to cough, deep breathe and ambulate effectively.

> Pulmonary pearls: early extubation; incentive spirometry, coughing and deep breathing; early ambulation; adequate pain control

PREVENTION OF INFECTION

Prevention of infection is key to ensuring optimal outcomes and quality of life for patients requiring mechanical circulatory support. Infection prophylaxis begins prior to VAD insertion. Preoperative urinary tract, pulmonary or catheter related infections are aggressively treated with intravenous antibiotics. If possible, VAD insertion is delayed until the course of therapy is completed. Time permitting, a thorough dental examination and necessary extractions are performed to eliminate what could be a source of postoperative bacteremia. Broad spectrum perioperative antibiotic coverage is provided. Postoperatively all lines and tubes are removed as soon as possible. Pulmonary toilet includes early extubation, incentive spirometry and chest physiotherapy. Early mobilization improves pulmonary toilet and facilitates rehabilitation efforts. Ideally, patients are out of bed within 24 hours of VAD insertion and are ambulating with assistance within 48 hours. Subsequently, the frequency of ambulation is increased to at least four times per day for the duration of the patient's hospital stay. Adequate nutrition is instituted on the first postoperative day either through enteral or parenteral means. If there is an ongoing need for venous access, lines are rotated to new sites every 3 days. Skin condition, especially over pressure points including heels, coccyx, elbows and occiput is monitored for breakdown and the patient's position changed every 2 hours until he is able to turn himself and/or get out of bed independently. Any areas of skin breakdown are aggressively treated. Chest and abdominal incisions are left open to air after the first 12-24 hours and monitored for erythema, warmth or drainage. Incisions are painted with povidone-iodine or 4% chlorhexidine gluconate three times daily if wound drainage develops. If the patient develops a fever, blood, urine and sputum are cultured. The patient is examined in an effort to identify sources of infection. The fever work-up should include unlikely or unexpected septic foci including the sinuses, oral cavity, gallbladder and perirectal region.

> Infection prevention pearls: early removal of invasive lines and catheters; pulmonary toilet; early ambulation; good nutrition; meticulous drive line/cannulae skin exit site care

DRIVE LINE/CANNULAE SKIN EXIT SITE CARE

The most obvious and concerning potential source of infection is the drive line and/or cannulae that exit the body through a skin wound. Unless tissue ingrowth occurs into the velour on the drive line/cannulae, the skin site represents an ongoing portal of entry for bacteria. The drive line or cannulae and any associated mediastinal or preperitoneal VAD pocket drains are meticulously assessed and dressed using sterile technique three times each day. Drain exit sites are treated in the same manner as the drive line or cannulae skin exit sites.

During dressing changes the wound(s) is monitored for amount and characteristics of any fluid, drainage, evidence of granulation tissue and tissue ingrowth onto the velour on the drive line or cannulae. The drive line/cannulae skin exit site dressing change routine developed at The University of Iowa can be used with any commercially available VAD. The list of supplies used for the dressing change are listed in Table 9.6. The procedure we employ for drive line/cannula exit site dressing changes is outlined in Table 9.7. This dressing change can be performed by one person. If a patient requires long-term VAD support he is eventually taught to perform his own drive line/cannulae care. If the patient's body habitus or the location of the exit site prevents the patient from performing the exit site care a responsible caregiver is identified and trained. If an alternate skin exit site care plan is employed it is important to remember that the Thoratec VAD (Thoratec Laboratories Corp., Pleasanton, CA) should never be exposed to acetone products as they can crack the blood pump housing.

In addition to meticulous dressing changes patients must take utmost care not to wet the drive line/cannulae exit site during bathing. Patients usually choose to sponge bathe. If the patient has an implantable VAD he may shower provided the drive line exit site is well covered. A hand-held showerhead should be used making sure the water spray is directed well away from the exit site.

> Drive line/cannulae skin exit site care pearls: sterile dressing change three times each day; keep water off the exit site; reliable caregivers; treat drain sites the same as drive line/cannulae exit sites

---

Table 9.6. Drive line/cannulae skin exit site dressing change supplies

Masks
Sterile small drapes
Sterile gloves (appropriate size for patient/caregiver)
Nonsterile gloves
1 sterile tongue blade
1-2 sterile ABD pads (Kendall Corp., Mansfield, MA)
2 packages sterile Kerlix gauze (Kendall Corp., Mansfield, MA)
3 packages sterile 4 x 4 gauze
1 package sterile split 4 x 4 gauze
Rubbing alcohol
Sterile saline (250 ml bottles)
Hydrogen peroxide
1% silver sulfadiazine cream
Montgomery straps (Dermicel; Johnson & Johnson Medical, Inc., Arlington, TX)
Sterile 20 ml syringe
Sterile container for saline/hydrogen peroxide

Table 9.7. Drive line/cannulae skin exit site dressing change procedure

1. Wash hands. Nurse, patient and anyone else in room apply masks. Apply nonsterile gloves and remove old dressing. Discard nonsterile gloves.
2. Place small sterile drape on clean work surface that has been wiped down with rubbing alcohol. Drop all dressing supplies (Table 9.6) onto sterile drape. Apply sterile gloves.
3. Mix sterile saline and 3% hydrogen peroxide 1:1 in a sterile container.
4. Place two sterile 4 x 4 gauze pads beneath drive line/cannulae skin exit site. Using sterile syringe draw up 20 ml of the half strength hydrogen peroxide solution.
5. Irrigate three times (total of 60 ml) around exit site. Change gauze as needed when saturated.
6. Moisten a sterile Kerlix gauze (Kendall Corp., Mansfield, MA) with half strength hydrogen peroxide solution and gently wipe around exit site. Without going back to exit site wipe any remaining fluid or blood from area. Blot dry with sterile gauze.
7. Apply thin layer of 1% silver sulfadiazine cream with sterile tongue blade around exit site.
8. Repeat steps 3-6 to drain sites if present.
9. Place sterile split 4 x 4 gauze under and around drive line/cannulae skin exit site. Fold one sterile gauze pad and place under split 4 x 4 to support drive line/cannulae. Cover with remaining sterile 4 x 4's. Repeat this step for drain sites if present.
10. Cover drive line/cannulae and drain exit sites with sterile ABD pad(s) (Kendall Corp., Mansfield, MA). Secure with Montgomery straps (Dermicel; Johnson & Johnson Medical, Inc., Arlington, TX).
11. Perform dressing changes every 8 hours. If dressing becomes saturated or loose between regularly scheduled wound care increase frequency of dressing changes as needed.
12. Dressing change procedure must be taught to a reliable caregiver for patients who will be discharged home with a VAD.

NUTRITION

A preoperative assessment by a dietitian documents the patient's current nutritional status, eating habits, metabolic needs and lifestyle habits surrounding nutritional intake. The patient is educated about lifestyle changes necessary to meet good nutritional guidelines. Objective measurements of nutritional status include serum total protein, albumin and prealbumin levels. Caloric and protein intake are monitored and undernourished patients are provided with appropriate dietary supplements.

Mechanical blood pumps are often inserted under urgent or emergent conditions in patients who may not have been practicing good eating habits or who are malnourished due to chronic heart failure. Enteral feedings are instituted within 24 hours of VAD insertion. If the patient has an adynamic ileus parenteral nutrition is employed to avoid abdominal distension, vomiting and the potential for aspiration. When oral feeds begin intake is monitored and supplemental enteral feedings discontinued only when caloric needs are met by the oral route. Implantable VADs are associated with early satiety. The patient will often tolerate six small meals per day rather than three large ones. Sodium and fluid restrictions are employed in the immediate postoperative period as interstitial fluid collection is exacerbated by the capillary leak associated with cardiopulmonary bypass. When endothelial integrity returns to normal, usually within a few days, sodium and fluid restrictions are relaxed. Food items brought from home are more satisfying than hospital food to patients who have been hospitalized for a prolonged period of time. A small refrigerator and microwave oven in the patient's room or accessible in the unit allows the patient to store and prepare food that may be more appealing to him than that which appears on the hospital menu. Alternatively, the dietitian may work with the patient to obtain food that he likes and will eat even though the food item may not appear on the hospital menu.

The postoperative goal is to achieve positive nitrogen balance, meet caloric needs and allow the patient to gain weight through good nutritional habits. The desired endpoint in the occasional obese patient is weight loss. The dietitian, patient and rehabilitation personnel must work together to develop a plan to balance caloric intake and exercise such that net gain or loss is achieved based on individual need.

> Nutrition pearls: enteral/parenteral nutrition until adequate oral intake; eat less but more often; early dietitian involvement

MEDICATION

Common postoperative medications are listed in Table 9.8. Depending upon VAD type, postoperative anticoagulation may consist of a heparin sodium infusion followed by oral warfarin sodium in the patient receiving long-term VAD support. Patients supported with the Thermo Cardiosystems VAD require aspirin alone. Postoperative pain is initially managed with intravenous morphine sulfate.

Table 9.8. Commonly used medications for patients receiving mechanical ventricular assistance

Anticoagulants
    Heparin sodium
    Warfarin sodium
Analgesics
    Narcotics
    Non-narcotics
Antibiotics
    Perioperative
    To treat a specific infection
Stress ulcer prophylaxis
Stool softener/laxative
Vitamin with iron
Cardiovascular indications
    Diuretic
    Antihypertensive agents
Medications for other chronic conditions
    Diabetes mellitus
    Gout
    Hypothyroidism
    Estrogen therapy

The patient is converted to an oral analgesic including hydromorphone, acetaminophen with codeine or acetaminophen alone when oral intake resumes. Nonsteroidal anti-inflammatory agents are avoided due to the potential for gastrointestinal irritation and adverse effect upon renal function.

Perioperative antibiotics are discontinued within 24-48 hours of surgery. Thereafter, antibiotics are only used to treat a documented septic focus. Stress ulcer prophylaxis with a histamine blocker is initially administered intravenously but converted to the oral route when the patient is able to take enteral feedings. If the patient has no history of gastroduodenal ulcers or gastroesophageal reflux this medication is discontinued after the initial postoperative period. Stool softeners are prescribed to prevent straining while the patient is on narcotic analgesics. Stool softeners and laxatives are discontinued at the patient's discretion. Vitamin supplements with iron are used until the patient's nutritional status and hemoglobin level normalize.

Beyond the immediate postoperative period heart failure medications are rarely required. Digoxin, vasodilators and diuretics are usually unnecessary. The patient with long-standing hypertension may require antihypertensive medication. Treatment of chronic conditions including diabetes mellitus, gout and hypothyroidism must continue after VAD placement.

Patients requiring long-term VAD support are taught to take their own medications and record dosages and timing of administration in a diary. The patient continues the diary following discharge from the hospital as a method of assuring proper medication administration. This practice also ensures that patients become used to documenting their daily mediations as this routine is expected following cardiac transplantation as well. Before discharging a VAD patient from the hospital make sure that he can obtain all medications and will not stop taking them due to lack of financial resources. This potentially disastrous situation must be recognized and addressed during the predischarge assessment by the nursing coordinator and/or social worker.

> Medication pearls: anticoagulation is essential; remember medications for chronic conditions; home medication documentation; be sure patient can obtain medications following hospital discharge

REHABILITATION

Physical rehabilitation after VAD placement varies dramatically depending on the size, portability and mobility allowed by the VAD system employed. Patients are moved out of bed as early as the first postoperative day provided they are hemodynamically stable. This may consist of sitting in a chair or ambulating a few steps in the room. Mobilization and ambulation need not wait until the patient is transferred out of the intensive care unit. The VAD drive line or cannulae are secured as much as possible to limit motion of the conduit with respect to the skin during ambulation. Patient's bodies differ as do the drive line/cannulae exit sites.

Thus, there are a variety of ways to immobilize these conduits. For drive lines exiting the mid to lower abdomen an elastic binder with a vertical slit for the drive line serves this purpose well. The drive line is fixed in position with overlapping Velcro tapes. The drive line should not be secured to the patient's thigh as movement of the leg while walking can cause motion at the skin:drive line interface. The paracorporeal Thoratec VAD can be supported with a modified fabric pouch suspended from a neck strap. The fabric pouch used to carry cardiac telemetry monitors serves this purpose well.

If a patient's mobility is limited by the VAD system or a complication that precludes ambulation, rehabilitation physical therapy can still be performed. These patients should receive active or passive range of motion exercises to prevent contractures and loss of mobility. This activity should begin early in the postoperative course **before** the patient has been in bed and immobile for a prolonged period of time. VAD patients who are discharged from the hospital continue to participate in outpatient cardiac rehabilitation in order to improve endurance and strength. Local rehabilitation programs are taught about the device and how to safely exercise these patients prior to institution of therapy.

> Rehabilitation pearls: early ambulation; secure drive line/cannulae to limit its motion; passive/active range of motion if unable to ambulate; local cardiac rehabilitation for the patient discharged from the hospital

9

PSYCHOSOCIAL ISSUES

VAD patients have a seemingly overwhelming array of psychosocial needs that must be addressed in both the inpatient and outpatient setting. If psychosocial needs are met the patient will be better able to adapt to and accept the device and lifestyle changes that accompany it. If a VAD is implanted acutely there may be little or no preparation time for the patient and his significant others. Even with the benefit of preoperative education patients and families are often overwhelmed and may not remember much of what is told to them due to the stress of critical illness, hypoperfusion or the amount and complex content of the information provided.

Healthcare personnel can intervene in many ways to assist patients and families in successfully dealing with the stress and uncertainty of having a VAD in place. Staff must not allow their own uncertainty or unfamiliarity with the device and associated care to be conveyed to the patient. Nurses are responsible for refreshing their own knowledge about the device when they know a patient will be coming to their unit. Consistency in what patients are told by nurses shift-to-shift, what they are taught about their device and how dressing changes are performed is essential in gaining their confidence and trust in the staff caring for them. During the inpatient stay help patients establish a daily routine and gain independence with their care as quickly as possible. This may be difficult on a busy unit when it is faster to provide care for the patient rather than teach them to care for themselves. If this is overlooked, however, the situation may develop where a patient is medically prepared for discharge but may not be able to correctly administer medications, care for the drive line or troubleshoot the device correctly if a problem occurs.

During the inpatient hospitalization, especially if it is prolonged, the patient should be allowed to assume responsibility for their care whenever possible. Be flexible when developing care plans and allow patients to participate in decision making as there are many aspects of their care over which they have no control (Table 9.9). Patients requiring an extended inpatient stay may feel disconnected from their families. Telemedicine, telephone calls and integration of the patient's "hospital family" with his nuclear family can improve the patient's connection with the world outside of the hospital. This can be achieved, for example, by organizing a monthly potluck dinner for the hospital staff, the patient and his family. Another service beneficial to the patient who has received the VAD as a bridge to cardiac transplantation is the transplant support group. This group exposes the patient and his family to information about mechanical circulatory support and cardiac transplantation that cannot be provided by the healthcare team. Meeting other patients who are waiting for transplantation, with or without a VAD or who have undergone transplantation, provides the patient with the "I have experienced it" perspective.

The patient and family will need resource information for lodging near the implant center. The information packet should describe facilities within the medical center and those available within the community. The family may then plan for lodging needs both during the inpatient hospitalization and any outpatient periods when the patient needs to be near the center.

Having a VAD in place makes patients and families fearful for many reasons.

Table 9.9. Ways in which to enhance patient control during a prolonged hospitalization

---

Allow the patient to plan his daily schedule.
Allow uninterrupted sleep during hours the patient chooses.
Permit flexible visitation by significant others.
Allow pets to visit (if not to the ward, to some other hospital location).
Arrange for the patient's hairdresser or massage therapist to come to the unit to provide services.
Educate the patient so he is able to take his own medications and document vital signs.
Allow food to be brought in from outside the hospital.
Provide microwave oven and small refrigerator for patient's food.
Allow personal items from home such as clothing, quilts, personal computers etc.
Provide privacy and allow intimacy.
Facilitate trips off the unit when possible.
Facilitate trips outside the hospital when possible.
Permit a significant other to stay with the patient and provide care.
Facilitate use of relaxation or imagery techniques.
Elicit services of other professionals to help meet patient's needs such as a pastor, social worker, activities therapist, psychiatrist, counselor, hospital volunteers or other patients.
Arrange for the patient to attend transplant support group.

They fear infections, prolonged hospitalizations, that they may never get off the device or, in the bridge to transplant scenario, that they may die before transplantation. Patients also fear device failure and learning emergency procedures such as dealing with alarm situations or manually pumping a device. It is essential to teach patients and their families to troubleshoot the device very early in the hospitalization and to approach these procedures calmly and matter-of-factly. The patient and his support persons are taught to respond to alarm situations and manually pump the VAD. It is very helpful to use a demonstration VAD so the patient and family can walk through each emergency procedure slowly step-by-step before actually performing it on the patient. The patient may become lightheaded when the VAD is being converted from the primary driver to the back-up mechanism. Thus, it is important that the patient be in a supine position and that a trained staff member supervise all sessions when these procedures are performed on the patient.

Once the bridge patient has recovered from the implant operation preparations are made for hospital discharge. The impact of a chronic illness will have a finite impact on the patient and family. Roles within the family will be in flux depending upon the patient's symptoms and extent of recovery. Depending upon the acuity of the patient's clinical condition at the time of VAD insertion, the patient's ability to perform acts of daily living and routine household tasks may change dramatically. If the patient presented with an acute infarct their functional abilities will be drastically reduced compared to the functional status before the hospitalization. Conversely, the patient with end stage heart failure will undoubtedly enjoy a markedly improved quality of life following VAD insertion. Regardless, the family dynamics will change during the period of VAD support. The ongoing changes in family dynamics can affect the patient's self esteem, coping and communication between the patient and other family members.

The economic impact of VAD insertion and a prolonged hospitalization is not to be underestimated. Healthcare costs can be exorbitant. The patient may no longer be able to work. Other family members may be forced to seek employment or the amount they are required to work may increase or decrease. Financial counseling is an important part of the patient's ongoing care. An assessment of how the patient and family are coping with the financial burden imposed by a chronic illness is performed on an ongoing basis. The patient's payor sources should be reviewed periodically and patients directed to report any changes in payor coverage. Educate the patient and family about potential resources for assistance. To reduce the financial burden and avoid disruption of work schedules minimize follow-up visits and coordinate appointments to reduce the number of trips the patient must make to the medical center.

To assist with coping, patients and families may benefit from ongoing or periodic mental health services. These services are provided at the implant center or through community providers. If the patient lives a significant distance from the hospital utilizing community providers is more cost effective and efficient for the patient. In this instance the implant center should educate the provider about mechanical ventricular assistance and its impact on the patient and family. Mental health services are provided by a psychiatric nurse, psychiatrist, social work or other qualified mental heath professional.

Meeting the psychosocial needs of the VAD patient and their family can be a challenging aspect of their care. A thorough preoperative assessment provides an opportunity for early intervention. Ongoing screening identifies problems that can be dealt with in a timely fashion. Stress reduction enhances the patient's sense of well-being.

> Psychosocial pearls: allow choices; promote early independence; teach emergency procedures early in the patient's hospitalization; develop a support network; address financial concerns

HOME DISCHARGE PREPARATION

Preparing the VAD patient for the return to their home and community is a challenging and rewarding experience for the nurse, patient/family and local community members. To make this transition both safe and successful home discharge preparation is done in an organized, stepwise fashion. To promote the best patient outcome four stages of education and training are recommended.

STAGE I (IN-HOSPITAL)

Time and patient condition permitting, the first stage of home discharge preparation should begin prior to VAD insertion. Education and training begins with the patient, primary support person and other willing family/friends selected by the patient. This is usually the most difficult phase as the patient and family are under a great deal of stress. To make patient and family education efficacious it helps if the primary educator for the VAD patient is knowledgeable about the basics of patient education. The three essential principles of patient education include an assessment of patient readiness to learn, selection of teaching tools and adult learning principles.

An assessment of readiness to learn is vital to a successful educational experience. Readiness describes evidence of motivation at a particular time and addresses motivation and willingness of the learner. Patients and/or family members may not be in a state of readiness for learning. This must be taken into consideration prior to beginning home discharge teaching. The nurse must select the appropriate teaching tool. To do so, the nurse must understand how the patient/family best learn new information. For example, do they learn best with written or visual material? Do they learn best by actual hands-on training? Most people are able to tell the instructor this information that, in turn, determines the teaching tools to be used. VAD patient educators should be familiar with and utilize teaching principles of the adult learner as they educate and prepare materials. The teaching principles are #1: create an atmosphere that promotes collaboration, mutual trust, respect and mutual support and #2: learners must participate in forming learning objectives, designing and carrying out learning plans and evaluating their learning.[2]

Education begins with an overview of how the VAD system works with particular emphasis on normal system operation. After normal system operation has been taught the focus switches to troubleshooting the system, recognizing problems, alarm conditions and appropriate interventions for emergencies. Teaching during this phase utilizes competency and equipment checklists to aid in organizing and documenting training that has been completed. The competency checklist incorporates a section that documents what has been taught and when the patient and family members return demonstrate it. The competency checklist is particularly helpful if there are a number of family members, support people or staff members involved in the education process.

During Stage I the patient is instructed to keep a daily diary to monitor his temperature, pulse rate, blood pressure, pump rate, volume, flow rate, daily/weekly pump maintenance activities and daily medications (Table 9.10). Once the patient has recovered from VAD implantation he maintains the diary for the duration of VAD support. The diary becomes particularly important once the patient is discharged from the hospital. The diary should be brought to each clinic visit

---

Table 9.10. Patient diary

DATE: _____
WEIGHT: _____
Time: _____   Time: _____
Pulse Rate: _____   Pulse Rate: _____
Blood Pressure: _____   Blood Pressure: _____
Temperature: _____   Temperature: _____

MEDICATIONS	AM	noon	PM	PM	bedtime

DAILY   WEEKLY

Time: _____   Rate: _____   Maintenance activities:
                    Volume: _____   _____
                    Flow: _____   _____
                                                _____
Time: _____   Rate: _____
                    Volume: _____
                    Flow: _____

NOTES

for review by healthcare providers. Prior to hospital discharge the patient and support person are given an identification card that includes information on how and when to contact the healthcare team (Table 9.11). A healthcare team member familiar with the VAD is available for patient calls 24 hours a day, 7 days a week. The patient is instructed to carry this card with them at all times.

STAGE II (DAY PASS)

Once the patient and at least one support person have successfully completed the in-hospital education and training and the healthcare team feels the patient is prepared physically and psychologically, the patient proceeds to the second stage. This stage consists of a 1 day trip out of the hospital. The patient must be accompanied by a trained support person. Patients may go out to lunch or dinner, spend the day at a local shopping mall, go to a movie or to a local museum. The day trip gives the patient and support person a chance to gain confidence away from the hospital staff. During this day trip the patient is able to practice activities that he must master prior to permanent discharge. These activities include getting in and out of a vehicle, proper stabilization of the device drive line and awareness of surroundings so that the drive line does not get pulled or caught on any passing objects. During this day trip the patient and family become accustomed to curiosity from members of the community who have not seen a person with a VAD.

STAGE III (OVERNIGHT)

Once the patient has completed a successful day trip out of the hospital the patient is released for a period not to exceed 3 days. This stage includes at least one overnight stay out of the hospital. If the patient lives close to the hospital he is allowed to go home and return to the institution for evaluation. If the patient lives a distance from the hospital arrangements are made for the patient to stay in a local hotel. Although this stage may seem cumbersome it provides the patient and his family the opportunity to simulate the actual homegoing stage but in a location

---

Table 9.11. How and when to contact your nursing care coordinator

---

If you need to contact the ventricular assist device coordinator on call, call the hospital operator at *(area code) telephone number* and ask for pager number _____ . Remember to stay on the line.

Contact the coordinator on call for:
Any alarm conditions
Fever
Any changes in vital signs
Drainage from drive line exit site (change in quantity or quality)
Increased pain, swelling or redness around drive line
Lightheadedness
Syncope
Any illness (sore throat, flu, etc.)
Changes in any medications
Any actual or suspected equipment malfunction

close to the hospital where help is immediately available. During this stage the patient and family are responsible for moving and setting up the homegoing VAD equipment and supplies.

STAGE IV (PERMANENT DISCHARGE)

Upon successful completion of the overnight stay the patient returns to the implant center where an evaluation for permanent discharge takes place. The patient is released from the hospital if clinically stable and no new problems have arisen during the out-of-hospital experience. Once the patient has been discharged from the hospital frequent telephone checks and close follow-up through the outpatient clinic will ensure patient safety and optimize the chance for a successful outcome (Table 9.12)

EDUCATING THE LOCAL COMMUNITY

As the patient approaches hospital discharge the focus of education switches to the patient's community.

LOCAL AMBULANCE/EMERGENCY MEDICAL SERVICES

The patient's local ambulance and/or emergency medical services (EMS) must be prepared to deal with a VAD emergency or capable of transferring a VAD patient from a local hospital to the implant center. This training is most readily accomplished by the VAD coordinator who provides an inservice at a local healthcare facility. This inservice is particularly effective if timed with the patient's hospital discharge. Having the patient present at the inservice introduces the patient to EMS personnel. The patient is reassured to know that there are knowledgeable healthcare professionals in his community who can help should a problem occur away from the implant center.

Table 9.12. Follow-up activities for the home discharge patient

---

Telephone checks (at least weekly).
Clinic visits (specify frequency). At each clinic visit:
    Obtain interval history
    Review patient diary
    Psychosocial assessment (coping, anxiety)
    Interval physical examination
    Inspect drive line exit site
    Laboratory studies
    Dietary/rehabilitation status
    Review readiness for cardiac transplantation
Periodic interventions
    Review education materials with patient/family
    Have patient demonstrate emergency procedures
    Monitor expiration date on equipment

Handouts that highlight step-by-step emergency procedures are distributed. The handout should provide a detailed description of how to manually pump the VAD. A 24 hour a day contact telephone number is provided. In addition to handouts, video and written materials concerning the particular VAD are given to the local EMS to keep for review. Training is repeated at least every 6 months, more often if necessary due to staff turnover.

LOCAL CARDIAC REHABILITATION STAFF

The patient is in need of ongoing cardiac rehabilitation after discharge from the implant hospital. Local rehabilitation staff need to understand device function and how to safely exercise a patient who is supported with a VAD. The local rehabilitation staff should learn how to stabilize the drive line. They must be aware that arrhythmias may occur during exercise and understand how to treat them in a patient supported with a blood pump. The local rehabilitation staff should be trained in emergency procedures and how to contact the implant center. A VAD center cardiac rehabilitation staff member may be willing to provide this inservice or the VAD coordinator may be able to cover this material while educating the local EMS.

NOTIFICATION OF LOCAL ELECTRIC COMPANY

When a VAD patient is discharged to home the patient's local electric company is notified. When contacting the local electric company include the patient's account number and state that the patient is being discharged from the hospital on life support and must be given priority should an electrical outage occur.

REFERENCES

1. Redman BK. Teaching: Theory and interpersonal techniques. In: Van Schaik T ed. The process of patient education. 7th ed. St. Louis: Mosby 1993: 118-139.
2. University of Southern California. Applications in continuing education for health professions. In: Knowles MS ed. Andragogy in action. Applying principles of adult learning. 1st ed. San Francisco: Jossey-Bass Publishers 1984: 297-335.

9

Complications of Mechanical Ventricular Assistance

Wayne E. Richenbacher

INTRODUCTION

Patients who require mechanical circulatory support are by definition critically ill. If the patient is to receive a blood pump for postcardiotomy cardiogenic shock the patient will have already undergone an open heart procedure. Ventricular assist device (VAD) implantation further prolongs the cardiopulmonary bypass (CPB) time. A patient who receives a VAD as a bridge to cardiac transplantation suffers from end stage congestive heart failure with the constellation of problems associated with chronic systemic hypoperfusion. The patient is also housed in an intensive care unit with invasive hemodynamic monitoring lines, catheters, and possibly, intraaortic balloon counterpulsation and/or mechanical ventilation. Thus, complications that occur during the period of mechanical ventricular assistance may be related to the patient's clinical condition prior to VAD insertion, the device implantation procedure or develop postoperatively either as a result of the patient's hospitalization or device malfunction. The purpose of this chapter is to identify some of the more common problems encountered in a patient who is supported with a mechanical blood pump. Diagnostic techniques are described and treatment options summarized. Complications that develop in the patient receiving mechanical circulatory support are broken down into two groups. Early complications occur in the operating room at the time of blood pump insertion or in the immediate postoperative period (Table 10.1). Late complications occur any time the patient is being supported with a VAD (Table 10.2).

EARLY COMPLICATIONS

INADEQUATE LEFT VENTRICULAR ASSIST DEVICE (LVAD) FLOW

The first problem that can be encountered in the patient receiving LVAD support is inadequate LVAD flow. This problem is identified as CPB is discontinued

10

Table 10.1. Early complications

Inadequate left ventricular assist device (LVAD) flow
> Hypovolemia
> Right ventricular dysfunction
> LVAD inflow cannula obstruction

Right ventricular failure
Patent foramen ovale with systemic desaturation
Hemorrhage
Thromboembolism

Table 10.2. Late complications

Thromboembolism
Infection
> Not related to the ventricular assist device (VAD)
> VAD related
>> Percutaneous drive line or cannula colonization
>> Pocket infection (implantable VADs)
>> Endovascular

Multisystem organ failure
Pump related
> Device malfunction
> Costal margin pain
> Gastrointestinal compression (implantable VADs)

Pump dependency

and LVAD flow initiated. Regardless of the indication for use or device inserted, the LVAD flow index should always exceed 2.0 L/min/m². If the LVAD flow index is less than 2.0 L/min/m² the patient will probably succumb from multisystem organ failure secondary to end organ hypoperfusion. The best way to determine the etiology of inadequate LVAD flow is to monitor both the left atrial and right atrial pressures as CPB is discontinued (Table 5.5). Low left and right atrial pressures are indicative of hypovolemia. Volume administration readily corrects the problem. Upon termination of CPB and initiation of left ventricular assistance, a bulging, poorly contractile right ventricular free wall and infundibulum, associated with inadequate LVAD filling, a low left atrial pressure and high central venous or right atrial pressure are indicative of right ventricular dysfunction. Treatment or more appropriately stated prevention, begins **prior** to LVAD insertion. The right heart must be adequately protected. If cardioplegic arrest is used the surgeon should ensure that an adequate dose enters the right coronary artery. If there is a right coronary artery stenosis, consideration should be given to concomitant right coronary artery bypass grafting. The free wall of the right ventricle is protected from the ambient room temperature with topical ice or cold pack application. If right ventricular dysfunction is present inotrope and pulmonary vasodilator administration often corrects the problem. If LVAD filling is suboptimal despite inotrope and pulmonary vasodilator therapy the patient requires mechanical right ventricular assistance.

Inadequate flow of blood into the LVAD is another cause of low LVAD flow. Inflow cannula obstruction is diagnosed by a low or mid range right atrial pressure and high left atrial pressure. These pressures indicate that the right heart is functioning satisfactorily and blood is able to move from the right heart to the left heart. However, when blood arrives in the left heart, it is unable to enter the LVAD. Native left ventricular failure results in systemic hypoperfusion and a high left-sided filling pressure, the reason the decision was made to provide mechanical left heart support in the first place. Transesophageal echocardiography provides confirmatory evidence for LVAD inflow cannula obstruction. With left atrial cannulation the inflow cannula is positioned within the body of the left atrium, preferably directed toward the central orifice of the mitral valve. The left atrium and ventricle should be reasonably well decompressed although left atrial inflow cannulation rarely results in total left heart decompression. Thus, some native left heart contraction exists and the patient's aortic valve may open with ventricular systole. LVAD inflow cannula obstruction occurs when a small diameter inflow cannula is selected or the left atrial cannula is positioned incorrectly. Use the largest inflow cannula available. For an average sized 70 kg patient a 36-40 F diameter inflow cannula will usually suffice. If a cannula of adequate size has been used the echocardiogram may show that a cannula inserted via the right superior pulmonary vein or the junction of this vein and the left atrial free wall has traversed the left atrium and become wedged in the left superior pulmonary vein. Repositioning an adequately sized left atrial inflow cannula usually results in satisfactory left heart decompression and adequate LVAD flow.

If the patient has received left heart support using an LVAD with a left ventricular apex cannula, the echocardiogram should demonstrate a well decompressed left atrium and ventricle with a brisk flow of blood into the apex cannula by doppler color imaging. In most instances the patient's native aortic valve no longer opens as all left ventricular blood is captured by the LVAD. The most common cause of inflow cannula obstruction is an impediment to flow of blood into the apex cannula caused by either the left ventricular free wall or ventricular septum. As it is difficult to reposition an apical cannula it is important to take the time to ensure that the initial apical cannula placement is correct. A brief period of cardioplegic arrest allows the heart to be elevated. As most cardiomyopathic patients have a dilated heart, the left ventricle is usually cavernous. With the heart elevated the surgeon should indent the apex to determine the location of the ventricular septum. The left ventricular free wall obstructs the flow of blood into the apical cannula when the diaphragmatic tunnel is positioned too far laterally. When left ventricular assistance is initiated the heart is decompressed and the patulous left ventricular free wall can "drape" across the orifice of the apex cannula. If the diaphragmatic tunnel and hence the inflow cannula are positioned medially, this potential cause of obstruction is eliminated (Fig. 6.4).

Even if LVAD flow is initially satisfactory it is important to monitor LVAD flow as the implant operation is completed and the patient moved to the intensive care unit. Pericardial closure, recommended in the patient being bridged to cardiac transplantation, can compress the heart and reposition the inlet cannula both of which can impair LVAD flow. Similarly, prolonged CPB time in the postcardiotomy cardiogenic shock patient can produce significant myocardial edema. When an edematous heart is further compressed by multiple VAD cannulae within the mediastinum sternal closure may reduce VAD filling. If either maneuver impairs VAD filling it is best to leave the pericardium open (although a pericardial membrane can be sewn anteriorly without cardiac compression) or stent the sternum open. If the sternum is stented open mediastinal drains should be placed beneath a sterile occlusive dressing. Lastly, postoperative bleeding can lead to cardiac chamber compression and inadequate VAD filling. If the patient has been in the intensive care unit with adequate VAD flow for a number of hours, when VAD flow suddenly declines an echocardiogram will often demonstrate clotted blood within the mediastinum. The hemopericardium can cause chamber compression and an inadequate flow of blood into the VAD. The patient should be returned to the operating room for sternal reexploration and mediastinal irrigation. Relief of the tamponade results in a prompt increase in VAD flow.

RIGHT VENTRICULAR FAILURE

Right ventricular failure is not truly a complication associated with mechanical circulatory support but can make management of the patient requiring left heart support very difficult. As a result, most discussions of problematic LVAD patients usually include a section on right ventricular failure. If, following the initiation of left ventricular assistance the patient has a high right atrial pressure ($> 15\text{-}20$ mm Hg), a low left atrial pressure ($< 5\text{-}10$ mm Hg) and an inadequate LVAD flow index (< 2.0 L/min/m^2), medical therapy for right ventricular dysfunction is initiated. Inotropes that have a pulmonary vasodilatory effect are added to the patient's drug regimen. Isoproterenol hydrochloride (0.01-0.20 µg/kg/min IV) and milrinone lactate (50 µg/kg IV loading dose, 0.375-0.75 µg/kg/min IV continuous infusion maintenance dose) effectively increase right ventricular contractility and reduce right ventricular afterload. Nitric oxide (20-60 ppm) is a powerful pulmonary vasodilator that has in many instances obviated the need for mechanical right ventricular assistance.[1] If the patient's right ventricular dysfunction is refractory to medical therapy, a right VAD (RVAD) is inserted.

If the indication for mechanical circulatory support is postcardiotomy cardiogenic shock, the choice of RVAD includes the non-Food and Drug Administration (FDA) approved centrifugal pump or the FDA approved Abiomed BVS 5000 Biventricular Support System (Abiomed, Inc., Danvers, MA) or Thoratec VAD System (Thoratec Laboratories Corp., Pleasanton, CA). Avoid a hybrid pump situation. A VAD manufactured by the same company should be used for both right and left heart support. If there is a high likelihood of recovery of right ventricular function it is simplest to use a centrifugal or Abiomed pump. These two devices employ smaller diameter cannulae that simplify cannulae placement within a crowded mediastinum. If there is a low likelihood of ventricular recovery and the patient may be a candidate for cardiac transplantation, the Thoratec VAD is the

VAD of choice as the device is designed for long-term use eliminating the difficult transition from a short-term to a long-term VAD as occurs in the bridge-to-bridge scenario. In the bridge to transplant patient supported with an implantable LVAD, the Thoratec VAD is the only device approved for long-term support of the right heart.

PATENT FORAMEN OVALE

This problem has been discussed in Chapter 5: Ventricular Assistance for Postcardiotomy Cardiogenic Shock and Chapter 8: Anesthesia for Patients with Ventricular Assist Devices. Suffice it to say that in the patient receiving left heart support an unidentified patent foramen ovale can result in acute arterial desaturation.[2] Arterial desaturation occurs when even mild right ventricular dysfunction leads to a rise in right atrial pressure and a right-to-left shunt of unoxygenated blood across the foramen into the decompressed left atrium. An intraoperative echocardiogram and bubble study following the initiation of left ventricular assistance will identify a patent foramen ovale (Fig. 8.3). A patent foramen ovale, no matter how small, should always be closed. If the diagnosis has not been made and a patient desaturates while in the intensive care unit, LVAD flow is reduced immediately to decrease the right-to-left shunt. An urgent echocardiogram and bubble study are performed as preparations are made for an immediate return to the operating room for foramen ovale closure.

HEMORRHAGE

Postoperative bleeding is multifactorial in origin. A preoperative coagulopathy related to the use of sodium warfarin, heparin sodium, aspirin, dipyridamole or anti-inflammatory drugs may be compounded by hepatic dysfunction and its attendant clotting abnormalities related to systemic hypoperfusion in the cardiomyopathic patient. Patients with postcardiotomy cardiogenic shock experience prolonged CPB times. The CPB machine-induced dilution of coagulation factors is made worse by CPB-induced platelet dysfunction. Upon completion of VAD insertion inadequate reversal of heparin sodium can increase the bleeding from surgical sites throughout the operative field.

Patient preparation begins prior to surgery. Clotting parameters including a prothrombin time, partial thromboplastin time and platelet count are determined as part of the patient's preoperative evaluation. Component therapy is used to correct any abnormalities prior to taking the patient to the operating room. Vitamin K is only administered to cardiomyopathic patients who will receive the Thermo Cardiosystems HeartMate VAD (Thermo Cardiosystems, Inc., Woburn, MA). Vitamin K is not administered to patients who are to receive either the Thoratec VAD or Novacor N-100 LVAD (Novacor Division, Baxter Healthcare Corp., Santa Ana, CA) as it will interfere with the required systemic anticoagulation with sodium warfarin therapy postoperatively.

Aprotinin is utilized in all patients who are to receive a VAD as a bridge to cardiac transplantation and in patients who may potentially require mechanical circulatory support for postcardiotomy cardiogenic shock (Table 5.3).[3] Patients

10

who are to undergo a high risk operation and who are at risk for postoperative ventricular dysfunction receive an expeditious operation to minimize CPB time. Meticulous surgical technique is always employed with emphasis placed upon achieving adequate intraoperative hemostasis. Topical hemostatic agents including absorbable gelatin sponges (8 x 12.5 cm Gelfoam sponge; Pharmacia & Upjohn Co., Kalamazoo, MI) soaked in topical thrombin (Thrombin-JMI; Gentrac, Inc., Middleton, WI) aid greatly in this regard. Heparin sodium is completely reversed with protamine sulfate upon termination of CPB. Blood product administration is dictated by a specific laboratory abnormality using leukodepleted blood in the patient who will subsequently undergo a cardiac transplant. Sternal closure helps tamponade marrow bleeding.

Postoperative hemorrhage is defined as mediastinal tube drainage exceeding 400 ml/hr for more than 4 hours or the need for sternal re-exploration secondary to tamponade or ongoing mediastinal bleeding. In the immediate postoperative period all abnormal coagulation studies are corrected with the appropriate blood product. Systemic anticoagulation, appropriate for the device employed, is not begun until mediastinal tube drainage falls below 50 ml/hr for several consecutive hours. Patients requiring mechanical ventricular assistance do not necessarily demonstrate the usual hemodynamic parameters associated with cardiac tamponade.[4] Tamponade in the VAD patient is usually manifest by a decline in VAD flow. The surgeon should have a low threshold for sternal reexploration in the patient with inadequate VAD flow.

THROMBOEMBOLISM

Patients who suffer a perioperative neurologic event usually do so as a result of intraoperative low systemic flow or perfusion pressure, dislodgement of a left ventricular thrombus or inadequate LVAD deairing. Patients who have suffered a transmural myocardial infarction or who have a very low ejection fraction secondary to a cardiomyopathy are prone to left ventricular thrombus formation. The degree to which this thrombus is organized and adherent to the trabeculae carneae within the left ventricle is variable. At the time of VAD insertion it is wise to avoid manipulation of the left ventricle until after CPB is initiated and the aortic crossclamp applied. If the left atrium is cannulated it may not be necessary to disturb the left ventricle at all. For left ventricular apex cannulation the endocardial surface of the left ventricle is carefully inspected to ensure that all thrombus is removed (Fig. 10.1).

Although not thromboembolic in nature a patient may also suffer a stroke at the time of VAD implantation as a result of low systemic flow or perfusion pressure or inadequate left heart and LVAD deairing. Specific LVAD deairing maneuvers are described in Chapter 11: Device Specific Considerations. Although transesophageal echocardiography cannot determine the adequacy of VAD deairing, echocardiography is used to aid in native left heart deairing. The usual cardiac deairing maneuvers are employed including inversion of the left atrial appendage, placing positive pressure on the endotracheal tube to clear the pulmonary veins and shaking the heart to dislodge trapped air within the native left

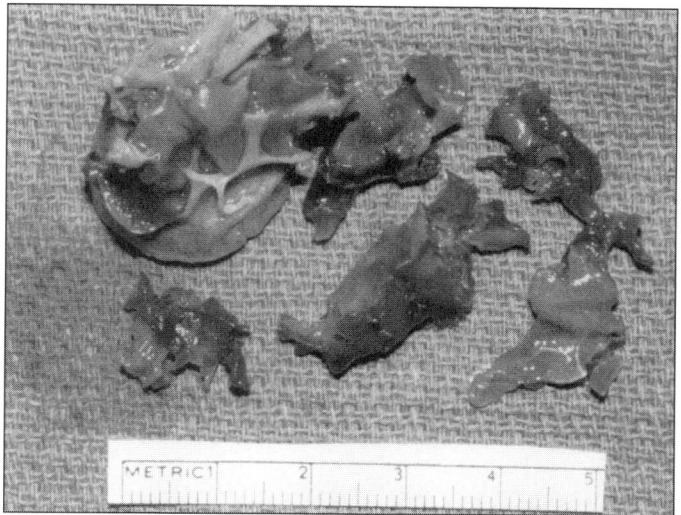

Fig. 10.1. This patient suffered an anterior wall myocardial infarction ten days prior to left ventricular assist device insertion. The myocardial core from the left ventricular apex is shown. Well organized thrombus is tightly adherent to the trabeculae carneae. Additional thrombotic material removed from the endocardial surface of the left ventricle is shown next to the core of myocardium.

10

atrium and ventricle. Deairing maneuvers are accomplished with the patient in Trendelenburg and continuous aspiration on an aortic root vent. The aortic root vent is not removed or the patient returned to a supine position until all air has been evacuated from the patient's heart and LVAD. As an added safety precaution, the LVAD is permitted to eject for several minutes prior to removing the aortic root vent and taking the patient out of the Trendelenburg position.

LATE COMPLICATIONS

THROMBOEMBOLISM
Late thromboembolic events, those which occur beyond the first postoperative day or two, are usually pump related. Late thromboembolism can result in a cerebrovascular accident with the associated neurologic deficit or end organ dysfunction as is the case, for example, with renal cortical infarcts. The thrombi originate at the blood:biomaterial interface in the VAD or inlet and outlet cannulae. The thrombi usually develop as a result of inadequate anticoagulation or a low flow state with blood stasis within the VAD.

Although control of postoperative bleeding, if any, is the primary concern immediately following VAD implantation, anticoagulation should begin as soon as mediastinal bleeding slows. We begin postoperative anticoagulation when the mediastinal tube drainage falls below 50 ml/h for several consecutive hours. Postoperative anticoagulation usually begins within 12-24 hours following surgery. For the Thermo Cardiosystems LVAD, aspirin (325 mg po or pr) is administered beginning on the first postoperative day. Patients supported with all other devices receive a continuous heparin sodium infusion at a rate to achieve an activated clotting time (ACT) of 180-200 seconds. Patients who receive a mechanical blood pump as a bridge to transplant receive sodium warfarin whenever they are able to take an oral diet. The goal is an international normalized ratio (INR) of 2.5-3.5. The heparin sodium infusion is discontinued when the INR exceeds 2.5. If a patient taking sodium warfarin requires an operative procedure during a period of VAD support, the patient is converted back to a heparin sodium infusion as the sodium warfarin is discontinued. The heparin sodium infusion is continued up to the time of the procedure, discontinued during the procedure and restarted as soon thereafter as possible.

VAD flow is maximized throughout the period of mechanical circulatory support. VAD flow is only reduced in the postcardiotomy cardiogenic shock patient while determining the degree to which ventricular recovery has occurred. Such weaning events are brief in duration and VAD flow is never reduced below 2.0 L/min. If a device component develops an adherent thrombus (within an extracorporeal VAD or its associated cannula connectors) and has to be changed, full systemic heparinization is used when VAD pumping is interrupted.[5] If an implantable VAD component must be changed, a formal operation is required and systemic heparinization is mandatory.[6]

Despite adequate anticoagulation thrombus and fibrin deposition can occur around any atrial cannula adjacent to the atrial wall. To avoid embolization of this thrombus/fibrin deposit, care is taken when any atrial cannula is removed. In the postcardiotomy cardiogenic shock patient who has return of ventricular function, positive pressure is placed on the endotracheal tube and the heart gently compressed as the left atrial cannula is withdrawn. As these maneuvers increase the left atrial blood is permitted to emanate from the left atrium prior to tying the surrounding pursestring sutures. Potential embolic material is ejected from the left atrium at that time. For the bridge to transplant patient the left atrial cannula should remain in place until CPB is initiated and the aortic crossclamp has been applied. The recipient cardiectomy is performed, the left atrial cannula withdrawn and the inside of the left atrium is inspected to ensure that all thrombus and fibrin have been removed.

INFECTION

A postoperative infection, regardless of etiology, can be a devastating complication if the responsible organism seeds the VAD.[7,8] VAD endocarditis can result in device malfunction or systemic embolization of vegetations. In the bridge to transplant patient, systemic sepsis may be a lethal event as the blood pump cannot be sterilized even with long-term antibiotic therapy. At worst, the infection may preclude transplantation. At best, the infection can result in a complicated posttransplant course in the immunosuppressed recipient. Infection prevention begins preoperatively. All invasive hemodynamic monitoring lines are removed and replaced with new lines through new insertion sites the night prior to surgery. A preoperative urinalysis, urine culture, chest roentgenogram, sputum gram stain and culture are obtained on all patients. If the patient is to undergo high risk revascularization or another operation that might result in postoperative ventricular dysfunction and the patient has an identifiable septic focus on the preoperative work-up, the operation should be delayed until the infection is treated. Unfortunately, the luxury of time is rarely available to the bridge to transplant patient. As the potential transplant recipient deteriorates and is prepared for blood pump insertion, septic foci should be identified and treated with intravenous antibiotics as early in the patient's hospital course as possible. Even though the patient's infection may not have been treated with a full course of antibiotics, adequate blood levels of the appropriate antibiotic may reduce the potential for postoperative septic complications.

At the time of blood pump insertion or a high risk operation, appropriate antibiotic coverage is provided. In the operating room strict asepsis is maintained. Traffic in and out of the operating room is minimized and the operation is performed in an operating room that contains an antibacterial air filtration system and is under positive pressure with at least 15 air changes per hour. A "sterile" area is maintained around the operating table to discourage nonscrubbed "interested observers" from leaning over the operative field. Upon completion of the operation hemostasis is achieved and adequate mediastinal drainage provided. A mediastinal blood collection is an excellent medium for bacterial colonization. Postoperatively the patient is extubated and all lines and tubes removed as soon as the patient's clinical condition permits. A postoperative infection is defined by fever, leukocytosis often accompanied by a left shift in the differential or positive body fluid culture requiring antibiotic therapy. Postoperative nonVAD, related infections including a urinary tract infection, pneumonia, phlebitis and sinusitis are treated as they would be in a patient who is not supported with a mechanical blood pump.

VAD-related infections can be further defined as drive line or cannula related, endovascular or pocket related in the case of implantable blood pumps. Preventive measures for VAD drive line or cannula infection begin in the operating room with device preparation and implantation. After the VAD is assembled it should be covered with an antibiotic soaked sponge (vancomycin hydrochloride 1 gm in 1000 ml of normal saline) until it is implanted. In particular, the portion of the VAD drive line or cannula that is velour covered should be soaked in the antibiotic containing solution or wrapped tightly with an antibiotic soaked sponge. The "wetability" of these surfaces is not particularly good and it is imperative that all interstices be exposed to the antibiotic solution prior to blood pump implantation. Drive line and cannulae skin exit sites should be snug. Avoid excising large skin buttons. If the skin and soft tissue lie tightly against the textured surface of the drive line or cannulae the potential for tissue ingrowth is maximized. Postoperatively, motion of the drive line or cannulae with respect to the skin exit site is minimized to further enhance the potential for tissue ingrowth and fixation of the drive line or cannulae in the subcutaneous tissues. Exit site care is directed by the manufacturer's recommendations. We use a regimen that is easy to perform, gentle on the adjacent skin, protective against most cutaneous organisms and not injurious to any drive line or cannula. Using sterile technique the drive line or cannula is cleaned with hydrogen peroxide (3% diluted 50:50 with normal saline), dried, coated with a thin layer of 1% silver sulfadiazine cream and covered with a sterile dressing every 8 hours (See Chapter 9: Nursing Care of the Patient Requiring Mechanical Circulatory Support).

Despite exemplary operative and postoperative management, percutaneous skin exit sites serve as an ongoing source of infection. Drive line or cannula infections are most often localized colonization of the skin:biomaterial interface and underlying soft tissue that rarely ascend into the VAD pocket or mediastinum. Colonization of the skin exit site is characterized by a purulent discharge adjacent to the drive line or cannula that may or may not be associated with signs of systemic sepsis. Skin exit site colonization can usually be controlled with local dressing changes and oral antibiotics. Long-term antibiotic therapy is usually unnecessary, and in fact, localized colonization is often "curable" with local measures alone.

VAD pocket or mediastinal infections, the latter in the case of extracorporeal VADs, may occur de novo or as a result of an ascending drive line or cannula infection. These infections present as fever, an elevation in the patient's white blood cell count and a VAD pocket or mediastinal fluid collection that will in later stages drain though the midline incision, drive line or cannula skin exit site. These infections usually present several weeks after VAD insertion and occur less commonly after the skin has "healed" to the textured drive line or cannula surface. The drainage or fluid collection should be cultured and appropriate intravenous antibiotic therapy instituted. The pocket or mediastinum must be surgically explored, irrigated and drained. We utilize a pulsed irrigation system (SurgiLav Plus Pulsed Irrigation System; Stryker Instruments, Kalamazoo, MI) with warm irrigation solution (neomycin sulfate-polymyxin B sulfate solution for irrigation 1 ml in 1000 ml normal saline). The pocket is drained with two 10 mm flat, fluted silicone drains (Blake drain; Johnson & Johnson Medical Inc., Arlington, TX) and irrigated continuously postoperatively (1 gm vancomycin hydrochloride in 1000 ml normal saline). The irrigation solution is administered through one drain at 20 ml/hr. The effluent is collected from the second drain and the output recorded and totaled. If this input exceeds the output by > 50 ml, the irrigation administration is discontinued until the output matches the input. Other described techniques for management of implantable VAD pocket infections include muscle flap or omental packing, redirection of the VAD drive line to a new percutaneous skin exit site and implantation of antibiotic impregnated methylmethacrylate beads.[9] VAD pocket infections are never "cured". The patient who develops such an infection should remain on intravenous antibiotics with VAD pocket irrigation until the VAD is explanted at the time of cardiac transplantation.

The most dreaded infectious complication of mechanical circulatory support is an endovascular infection usually of the blood contacting VAD surface. These infections are blood-borne and classically caused by staphylococcus or fungus. The patients develop fever, leukocytosis and positive blood cultures but may also present with acute VAD valvular regurgitation or a vegetation related thromboembolic event. Transesophageal echocardiography aids in diagnosis. Vegetations may be identified at the endocardial-VAD apical conduit interface. Acute VAD inflow valve regurgitation is evidenced by a retrograde flow of blood into the native left ventricle during VAD systole; an echocardiographic finding very similar to native aortic insufficiency. VAD endocarditis rarely occurs in the postcardiotomy cardiogenic shock patient as the duration of VAD support is brief. Such a devastating infection can never be cured and the infected patient requires long-term antibiotic therapy. If the patient is still considered a candidate for cardiac transplantation consideration must be given to pump and/or inlet and outlet valved conduit replacement. If the patient is supported with the extracorporal Thoratec VAD, pump replacement can usually be accomplished without reopening the patient's incision.[5] Special attention must be directed to VAD deairing as the prosthetic vascular graft portion of the outlet cannula is not exposed. Replacement of an implantable VAD, in particular the left ventricular apical inflow conduit, can be a daunting surgical undertaking.[6] Exchange of an implantable VAD requires femoral-femoral CPB and cardioplegic or profound hypothermic circulatory arrest. Exposure of the left ventricular apex usually requires that the left hemidiaphragm be taken down from the costal margin. The VAD drive line is transected within the VAD pocket and withdrawn from the subcutaneous tunnel. The VAD itself is removed from the operative field. Vegetations are sent for gram stain and culture. The new VAD is inserted by creating a new drive line skin exit site and connecting the VAD to the old apical and outlet graft connectors.

MULTISYSTEM ORGAN FAILURE

Patients who require mechanical circulatory support usually have preexisting end organ dysfunction. In the postcardiotomy cardiogenic shock patient this may be related to pre- or intraoperative hypotension or hypoperfusion. Hemoglobinuria secondary to a prolonged CPB time exacerbates any underlying renal dysfunction. Intraaortic balloon counterpulsation may further interfere with intestinal and renal perfusion. In the bridge to transplant patient chronic congestive heart failure and the associated low cardiac output is manifest by hepatic dysfunction and prerenal azotemia.[10] End organ function is further impaired in the patient with central venous hypertension secondary to right heart failure. Fortunately, the incidence of multisystem organ failure following VAD insertion has declined as patient selection has improved. VAD insertion is not a salvage procedure. As results of mechanical circulatory support have improved over time patients have been referred for and undergone VAD insertion at an earlier point in the time course of their disease state. End organ function has not deteriorated to the point where organ dysfunction is irreversible. Provided the patient undergoes VAD implantation in a timely fashion, one of the primary determinants of survival thereafter is the adequacy of LVAD flow. Marginal systemic perfusion (LVAD flow index < 2.0 L/min/m²) is associated with an ongoing decline in end organ function.

End organ function in patients who receive VAD support at an early point in the disease process, who have adequate systemic perfusion following LVAD insertion and who suffer no infectious complications usually normalizes in 2-4 weeks (Fig 10.2). The timing of VAD explantation in the postcardiotomy cardiogenic shock patient is determined by myocardial recovery. As myocardial recovery usually occurs within 7-10 days, the VAD may be removed before renal and hepatic function have returned to normal. This consideration justifies a delay in VAD removal in this patient population until myocardial recovery is complete. This management approach precludes the need for alpha agents that may adversely affect end organ perfusion following VAD explantation. In the bridge to transplant patient scenario, the patient is not relisted for cardiac transplantation until end organ function, nutritional indices and rehabilitation parameters have normalized. This management approach helps ensure a good outcome following transplantation. A late decline in end organ function is usually related to occult sepsis or thromboembolism. The usual laboratory studies and imaging techniques help elucidate the source of sepsis or end organs affected by thromboembolism.

PUMP RELATED
Device malfunctions are rare, a tribute to extensive preclinical device testing and experience gained during clinical trials performed as a part of the regulatory affairs process required by device manufacturers to achieve FDA approval for a

10

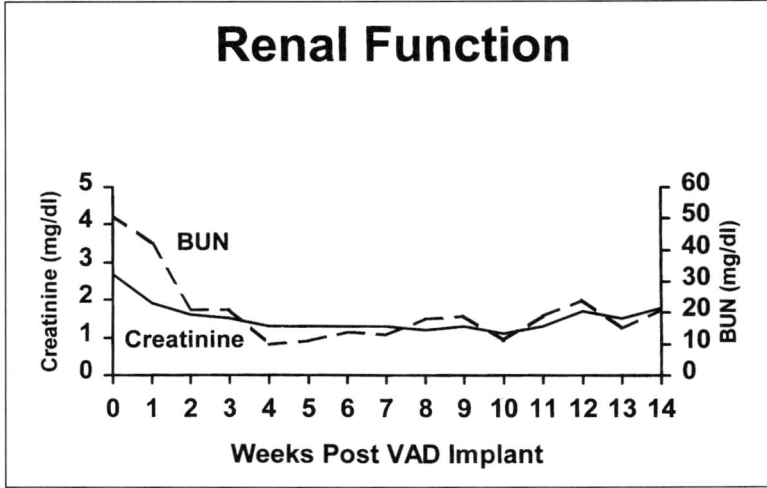

Fig. 10.2. Time course of recovery of renal function in a patient requiring long-term left ventricular assistance as a bridge to cardiac transplantation. BUN = serum urea nitrogen, VAD = ventricular assist device.

VAD. Device malfunctions, either minor or major, are heralded by controller or drive unit alarms. The response to an alarm situation is specific and well outlined in the clinician and patient users manuals provided by the VAD manufacturer. Suffice it to say that these devices are designed with a great deal of redundancy. In the event of a device/power source failure the device will usually continue to function in some sort of back-up mode. Should the device/power source fail altogether most devices can be hand-cranked or hand-pumped until the primary problem is rectified. A catastrophic cessation in device function is a rarity.

Besides thromboembolism there are a number of problems that are VAD related. If the VAD cannulae or the pump portion of an implantable VAD impacts upon the costal margin, the patient can experience chronic pain particularly when rising from a supine position or when assuming a sitting position. This is particularly true when an implantable pump is placed in the preperitoneal position in a patient with a small body habitus. The best treatment is avoidance. Cannulae for extracorporeal VADs should exit the skin at least two fingerbreadths below the costal margin. This will not only lengthen the subcutaneous tunnel but also move the cannulae away from the costal cartilage. If an implantable pump is placed in the preperitoneal location the VAD pocket is developed by taking the peritoneum down from the caudad surface of the diaphragm. The diaphragmatic tunnel for the left ventricular apex cannula should be positioned well away from the costal margin. This decreases the chance of the pump impacting upon the costal margin and more importantly ensures that the apex cannula traverses the diaphragm directly into the left ventricular apex. An apical cannula placed too far ventrally is prone to obstruction either as a result of a cannula kink or malposition within the left ventricular apex. Although intercostal nerve blocks may afford the patient some relief, costal margin pain is usually mechanical rather than neurologic in origin. Thus, costal margin discomfort can rarely be relieved entirely.

An implantable VAD, whether placed intraperitoneally or in a preperitoneal position, can cause gastric compression in the patient with a small body habitus (Fig. 10.4).[11] Gastric compression can cause early satiety. Nutritional repletion is still possible using a postgastrectomy type diet with multiple small feedings spread throughout the day. Implantable VADs located in the intraabdominal position have been associated with visceral perforation while VADs placed in the preperitoneal position may develop a pocket seroma if inadequately drained. Patients who receive VAD support as a bridge to cardiac transplantation are also at risk for the development of anti-HLA antibodies during the period of support.[12] Sensitization of cardiac allograft recipients is considered a risk factor for acute rejection during the early posttransplant period, as well as death from acute or chronic rejection. The rise in panel reactive antibody (PRA) levels during VAD support is thought to be related to blood product, in particular, platelet transfusion. Leukodepletion of blood products attenuates this effect. However, the blood contacting VAD surface may serve as an ongoing pro-inflammatory stimulus. The latter conclusion was proposed when it was observed that the PRA increased during VAD support in patients who received fresh frozen plasma alone at the time of blood pump insertion.[13] In bridge to transplant patients blood product usage is

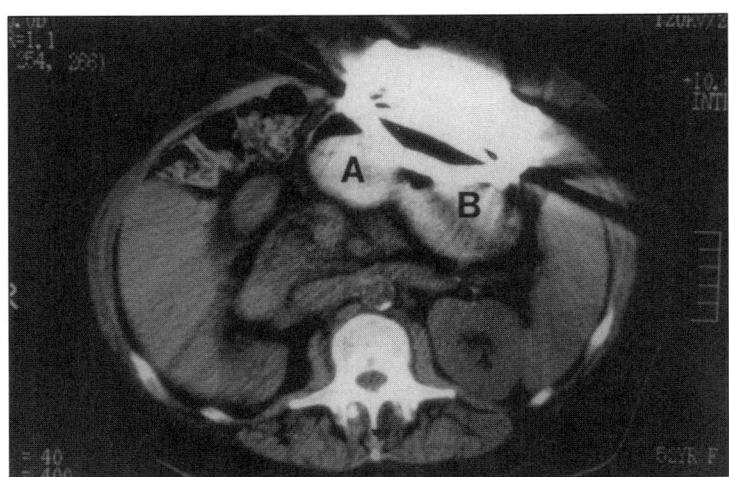

Fig. 10.3. Abdominal computed tomography of a small patient with an implantable left ventricular assist device. Enteric contrast fills the stomach. The antrum (A) and body (B) of the stomach are compressed by the convex surface of the blood pump.

minimized at the time of VAD insertion. If administered, packed red blood cells and platelet packs are leukodepleted. PRA levels are determined weekly for the duration of VAD support. PRA levels can rise months after blood pump implantation. If a patient develops a positive PRA, a prospective crossmatch is performed prior to cardiac transplantation. Pre- or perioperative plasmapheresis may lower the levels of anti-HLA antibodies before transplantation.

PUMP DEPENDENCY

In the postcardiotomy cardiogenic shock application, pump dependency occurs when the patient fails to recover myocardial function. This situation is best avoided by proper patient selection. A VAD should never be implanted with the expectation that ventricular function will improve in a patient who had a low preoperative ejection fraction or technically incomplete open heart operation. In other words, patient selection for VAD insertion for postcardiotomy cardiogenic shock should include patients who have a reasonable chance to recover function and be weaned from the device. If the patient has undergone a technically complete operation, had reasonable ventricular function to begin with and shows no sign of myocardial recovery after several days of VAD support, a cardiac transplant evaluation is initiated. If the patient is deemed a suitable candidate for cardiac transplantation and remains device dependent more than 10 days after VAD implantation the patient is taken to the operating room where the short-term VAD is removed and a long-term blood pump inserted. If the patient is not considered to be a candidate for cardiac transplantation the VAD may be removed and an attempt made to manage the patient with medical therapy alone. Such patients usually succumb to a complication of VAD support prior to having to make a decision to withdraw VAD support.

In the bridge to transplant patient, pump dependency occurs when the patient develops a complication of VAD support that precludes transplantation. The most common example is a stroke from which the patient cannot be rehabilitated. If the patient is septic but the infection suppressed with antibiotics the best option is still to proceed with cardiac transplantation.[7,8] In this scenario, transplantation is the best treatment option as the infection cannot be adequately treated until the biomaterial is removed. If the process that precludes transplantation is not fatal, a management option is VAD explantation. If the patient has been supported with the Thoratec VAD using left atrial cannulation, device explantation is accomplished without the use of CPB. The inlet cannula is withdrawn and the surrounding pursestring sutures tied. The outlet graft is transected adjacent to the aorta and the stump oversewn.

For the patient supported with a device using a left ventricular apex inlet cannula, be it the Thoratec VAD or an implantable blood pump, device explantation is somewhat more difficult. Femoral-femoral CPB is instituted and device removal accomplished under hypothermic, cardioplegic arrest. The outlet graft is transected adjacent to the aorta and the stump oversewn. If the device or mediastinum is infected, minimize the amount of biomaterial remaining behind by removing the outlet graft altogether. The defect in the ascending aorta is repaired primarily or with a bovine pericardial patch. The left ventricular apex is exposed and the apical cannula removed. The left ventricular apex is closed using a technique similar to that employed for left ventricular aneurysm resection.

10

REFERENCES

1. Argenziano M, Choudhri AF, Moazami N et al. Randomized, double-blind trial of inhaled nitric oxide in LVAD recipients with pulmonary hypertension. Ann Thorac Surg 1998; 65:340-345.

2. Magovern JA, Pae WE Jr, Richenbacher WE et al. The importance of a patent foramen ovale in left ventricular assist pumping. Trans Am Soc Artif Intern Organs 1986; 32:449-453.

3. Goldstein DJ, Seldomridge JA, Chen JM et al. Use of aprotinin in LVAD recipients reduces blood loss, blood use, and perioperative mortality. Ann Thorac Surg 1995; 59:1063-1068.

4. Smart K, Jett GK. Late tamponade with mechanical circulatory support. Ann Thorac Surg 1998; 66:2027-2028.

5. Lohmann DP, McBride LR, Pennington DG et al. Replacement of paracorporeal ventricular assist devices. Ann Thorac Surg 1992; 54:1226-1227.

6. Skinner JL, Bourge RC, Kirklin JK et al. Replacement of an intracorporeal left ventricular assist device. Ann Thorac Surg 1997; 64:839-841.

7. Fischer SA, Trenholme GM, Costanzo MR et al. Infectious complications in left ventricular assist device recipients. Clin Infect Dis 1997; 24:18-23.

8. Prendergast TW, Todd BA, Beyer AJ III et al. Management of left ventricular assist device infection with heart transplantation. Ann Thorac Surg 1997; 64:142-147.

9.   McKellar SH, Marks JD, Long JW. Treatment of heart-assist device infection with antibiotic-impregnated methacrylate beads. ASAIO J 1998; 44:36A.

10. Reinhartz O, Farrar DJ, Hershon JH et al. Importance of preoperative liver function as a predictor of survival in patients supported with Thoratec ventricular assist devices as a bridge to transplantation. J Thorac Cardiovasc Surg 1998; 116:633-640.

11. El-Amir NG, Gardocki M, Levin HR et al. Gastrointestinal consequences of left ventricular assist device placement. ASAIO J 1996; 42:150-153.

12. Massad MG, Cook DJ, Schmitt SK et al. Factors influencing HLA sensitization in implantable LVAD recipients. Ann Thorac Surg 1997; 64:1120-1125.

13: Stringham JC, Bull DA, Fuller TC et al. Avoidance of cellular blood product transfusions in LVAD recipients does not prevent HLA allosensitization. J Heart Lung Transplant 1999; 18:160-165.

10

Device-Specific Considerations

Wayne E. Richenbacher, Shawn L. Jensen, Scott D. Niles, James M. Ploessl

INTRODUCTION

The purpose of this chapter is not to rewrite the users manual provided by the blood pump manufacturer nor to serve as a substitute for the training programs provided by industrial representatives as part of the initiation of a mechanical circulatory support program. It is assumed that everyone involved with ventricular assist device (VAD) implantation and caring for patients who are supported with a VAD will have the opportunity to attend a training course provided by the device manufacturer or participate in inservice activities at the hospital in which they are employed. Rather, this chapter serves to provide the reader with a proven method of device assembly and technique for implantation. Operative and post-operative management protocols are described elsewhere in this text. Recognizing that there are multiple ways in which to approach a surgical problem, this chapter describes one or two reproducible implantation techniques for each of the devices that are commercially available.

11

CENTRIFUGAL PUMP

DEVICE PREPARATION

Unlike the other devices that are approved by the Food and Drug Administration for use as a VAD and as a result come packaged for use as such, centrifugal pumps and associated tubing and cannulae must be assembled from supplies available to the perfusionists. We use the Medtronic Bio-Medicus Bio-Pump (Medtronic Bio-Medicus, Inc., Eden Prairie, MN). As this pump is used for left heart bypass for descending thoracic aortic surgery the manufacturer has prepared a pump and tubing pack that can be readily adapted to allow the pump to be used as a VAD. The tubing pack contains a pump head, inlet and outlet tubing and flow probe all of which have been assembled to facilitate circuit priming (CBCCS Table pack-long with pump heads; Medtronic Cardiac Surgery, Medtronic, Inc., Ana-heim, CA). The circuit contains an exposed pump head and adjacent inlet and outlet tubing (Fig. 11.1). The next 170 cm of tubing are coiled into a sterile pack that is passed to the surgeon after the pump has been primed. The terminal ends of the inlet and outlet tubes exit the sterile wrap and have spikes on the ends to ensure sterile circuit priming. The inlet and outlet tubing are marked in several locations with blue and red tape, respectively.

The circuit is primed by clamping the inlet (blue) tube with the integral clamp and spiking a 1 L bag of Plasma-Lyte A (Baxter Healthcare Corp., Deerfield, IL). The clamp is released and the priming fluid permitted to run through the inlet tube in the sterile pack until it reaches the pump head. The pump head is held on its side to permit air to exit from the outlet port. When the priming fluid exits the pump head via the outlet port the outlet tubing is clamped between the pump head and flow probe with a tubing clamp. The tubing pack and pump head are shaken to ensure that any air remaining in the circuit exits the outlet port. The tubing clamp on the outlet tubing is moved below the air-prime fluid level adja-cent to the pump head. The pump head is placed on the drive console and the drive unit turned on at 1000-1300 rpm. The tubing clamp on the outlet side of the pump head is released and priming fluid permitted to pass through the remainder of the outlet tube. When all gross air has been removed from the circuit the distal end of the outlet tube is clamped with the integral clamp and the outlet spike inserted into the second port on the Plasma-Lyte A bag. The clamp on the outlet tube is removed and priming fluid permitted to recirculate. Recirculation contin-ues until such time as the inlet and outlet cannulae have been placed in the patient and are ready to be connected to the remainder of the circuit.

11

Fig. 11.1. The priming circuit for the Bio-Pump centrifugal pump.

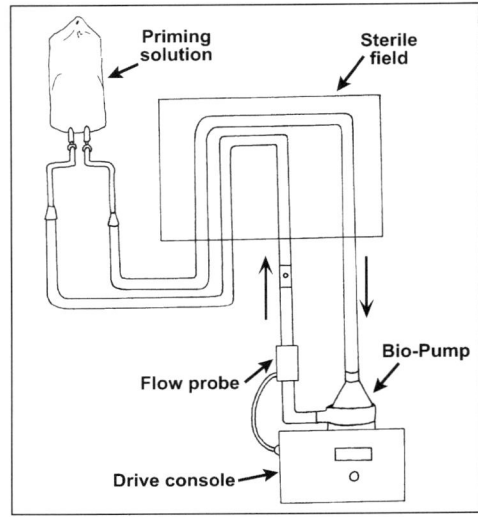

DEVICE IMPLANTATION

The centrifugal pump can be used with any uptake and arterial cannulae. We find the cannulae produced by Research Medical, Inc., (a Division of Baxter Healthcare Corp., Midvale, UT) to be ideally suited for use as VAD cannulae (Table 11.1).[1] The inlet cannulae are designed with a lighthouse tip, are wire reinforced and long enough to permit extracorporeal connection to the remainder of the VAD circuit. We prefer the 120° angle inflow cannula as it directs the cannula tip toward the atrioventricular valve reducing the potential for inflow cannula obstruction. We also prefer the 20° angle outflow cannula as the cannula lies along the anterior mediastinum while the tip is directed into the ascending aorta/aortic arch or main pulmonary artery depending upon which ventricle is being supported.

The cannulae are inserted through pursestring sutures as described in Chapter 5: Mechanical Circulatory Support for Postcardiotomy Cardiogenic Shock. The cannulae exit the chest in the subcostal region. In preparation for connection to the VAD circuit the cannulae are deaired and the exteriorized nonwire wound portion clamped with a tubing clamp. At this point the perfusionists unwrap the table pack (containing the coils of sterile inlet and outlet tubing). The surgeon should clamp the distal ends of the lines 3-5 cm into the sterile area leaving the colored tapes intact to aid in orientation as the cannulae are attached to the circuit. The cannulae are connected to the circuit tubing with $3/8$"-$3/8$" straight connectors. Although deairing of the connectors is made easier by the presence of a luer fitting, we believe this serves as a potential site for thrombus or fibrin deposition and thus prefer to use nonluer connectors. Although the tubes are marked with colored tapes it is always wise to check the pump head when attaching the

Table 11.1. Research Medical cannulae* used for centrifugal pump ventricular assistance

Inflow (all with lighthouse tip)		
32 F	120° angle	50 cm long
36 F	120° angle	50 cm long
32 F	90° angle	50 cm long
36 F	90° angle	50 cm long
40 F	90° angle	50 cm long
Outflow		
22 F	20° angle	50 cm long
24 F	20° angle	50 cm long
24 F	Straight	60 cm long

* Manufactured by Research Medical, Inc., a Division of Baxter Healthcare Corp., Midvale, UT.

11

cannula to the circuit tubing to ensure that orientation is correct. As the pump head was deaired at the time the circuit was primed, VAD flow can be initiated and cardiopulmonary bypass (CPB) flow reduced once the circuit is completed.

ABIOMED BVS 5000 BIVENTRICULAR SUPPORT SYSTEM

DEVICE PREPARATION

The Abiomed VAD (Abiomed, Inc., Danvers, MA) comes prepackaged with inlet and outlet tubing and integral cannulae connectors (Fig. 11.2). The cannulae connectors are covered with a latex adapter the free end of which possesses a straight three-eighths inch connector. The entire set is passed onto the sterile field. Two additional lengths of sterile tubing (three-eighths inch ID, 1.5 m long) are passed onto the sterile field. One length of tubing is attached to the blood pump inflow (atrial/blue) connector. The second length of tubing is attached to the outlet (arterial/red) connector. The latex tubing and associated connectors are fixed to the drapes on the operative field. The blood pump and open ended sterile tubing are passed to the perfusionist. The blood pump inflow tubing is attached to the outflow port (base) of an empty cardiotomy reservoir. A tubing clamp is applied to the tubing adjacent to the base of the cardiotomy reservoir. The blood pump outflow tubing is attached to the inflow port (top) of the cardiotomy reservoir. The cardiotomy reservoir is filled with 2 L of Plasma-Lyte A solution. The blood pump

Fig. 11.2. The priming circuit for the Abiomed ventricular assist device.

11

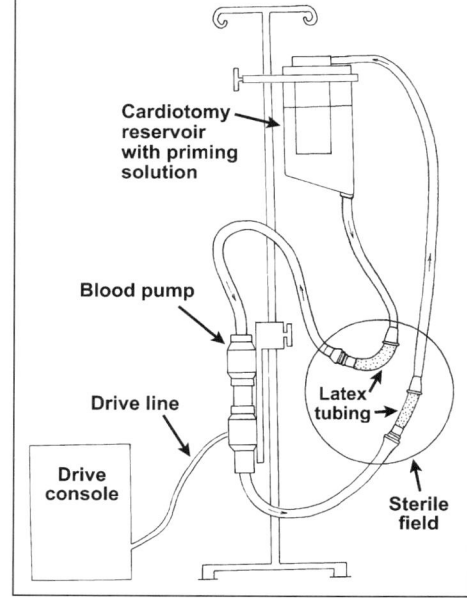

Cardiotomy reservoir with priming solution

Blood pump

Drive line

Drive console

Latex tubing

Sterile field

is inverted and elevated above the level of the cardiotomy reservoir. The tubing clamp is removed from the blood pump inflow tubing at the base of the cardiotomy reservoir. The inverted blood pump is slowly lowered allowing the prime solution to fill the inflow tubing and blood pump in an antegrade fashion. The blood pump is gently shaken to release air bubbles that tend to become trapped by the trileaflet valves. The prime solution exits the blood pump and returns to the cardiotomy reservoir via the outflow tubing. The blood pump is returned to its normal operating position (chamber with drive line down), connected to the drive console and activated. Priming is completed by allowing the prime solution to recirculate through the cardiotomy reservoir for at least 5 minutes. The blood pump and associated tubing are inspected to ensure that deairing is complete. Priming is discontinued by clamping the inflow and outflow tubing adjacent to the latex connectors. Tubing clamps are also applied to the two latex connectors. The latex adaptors are rolled off the cannulae connectors in preparation for connection to the inlet and outlet cannulae. The latex connectors and tubing to the cardiotomy reservoir should remain in the sterile field in case a second blood pump has to be primed. The inlet and outlet cannulae come in a variety of sizes (Table 11.2). As the device is gravity filled VAD flow is optimized by selecting the largest atrial cannula that will fit into the patient. The outlet graft is coated and does not require preclotting.

DEVICE IMPLANTATION

The atrial cannula is inserted in a manner similar to that used for centrifugal pump inflow cannulation. The inflow cannula is inserted through pursestring sutures. The malleable cannula works well if the dome of the left atrium is to serve as the insertion site.[2] The cannulae skin exit sites are located in the subcostal region. The atrial cannula is exteriorized. The tunneling bullet is placed in the end of the cannula. A heavy silk suture is passed through the eye at the end of the tunneling bullet. A Pean clamp (Pean artery forcep; V. Mueller, Deerfield, IL) is passed through the skin exit site into the pericardium. The silk suture is grasped with the clamp and the cannula withdrawn through the tunnel. The inlet cannula is inserted into the atrium after it has been passed through the tunnel. The cannula should be inserted at least 2 cm into the atrium. Depth marks (black lines) are located 1, 2 and 3 cm away from the end of the cannula.

*Table 11.2. Abiomed BVS 5000 Biventricular Support System cannulae**

Inflow		
46 F	90° angle	lighthouse tip
36 F	malleable	open tip
Outflow		
42 F	12 mm graft	
46 F	14 mm graft	

* Manufactured by Abiomed, Inc., Danvers, MA.

The outlet graft is trimmed to the appropriate length and anastomosed to the ascending aorta (or pulmonary artery for right ventricular assistance) with a 4-0 Prolene suture (Ethicon, Inc., Somerville, NJ). The anastomosis is checked for hemostasis before the cannula is exteriorized.

The inlet cannula is filled by momentarily compressing the heart or performing a Valsalva maneuver. A tubing clamp is applied to the nonwire reinforced exteriorized portion of the cannula. The arterial cannula is filled and the nonwire reinforced portion clamped. Cannula restraints (collets) are slipped onto both cannulae with the notched ends pointing away from the heart. The cannulae are attached to the blood pump tubing connectors. Deairing is accomplished by making these connections under a steady stream of saline. The cannula restraint is forced onto the blood pump tubing connector and fixed in position with the tie wrap provided in the accessory pack (Fig. 11.3). A final check is made to ensure deairing is complete before pumping is initiated.

THORATEC VAD SYSTEM

DEVICE PREPARATION

The Thoratec VAD (Thoratec Laboratories Corp., Pleasanton, CA) and associated electric and pneumatic leads are passed onto the sterile field. The Y-connector on the VAD is protected with a glove finger. The finger is placed over the Y-connector and tied in place to protect the connector from fluids. The VAD is supplied with a black collet nut and collet on the inlet side of the blood pump. If an atrial cannula is to be used for inflow, nothing need be done with the inflow

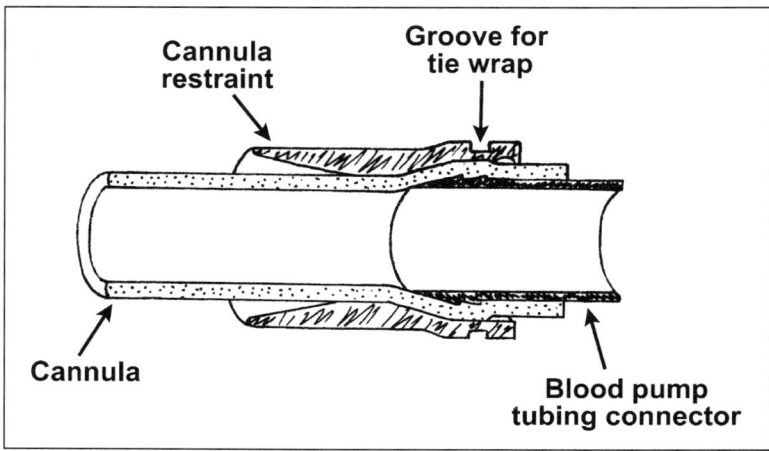

Fig. 11.3. The Abiomed cannula-to-pump connector using the cannula restraint.

collet nut and collet. If, however, a ventricular apex cannula is to be used the black collet nut and collet must be removed from the VAD inflow connector and discarded. The white collet nut and collet provided with the ventricular apex cannula are attached to the inflow connector on the VAD. One bottle of albumin (250 ml, 5% solution; Baxter Healthcare Corp., Glendale, CA) is mixed with 100 U of heparin sodium and the solution used to fill the VAD. This solution should remain in the VAD for a minimum of 15 minutes to ensure adequate passivation of the blood contacting surfaces.

Inlet and outlet cannulae are provided in a variety of sizes and configurations (Table 11.3). The atrial cannula does not require any specific preparation. If atrial cannulation is to be employed the size of the patient determines the length of the cannula ensuring that 1-2 cm of velour extend beyond the cannula-skin interface. If left ventricular assistance is to employ a left ventricular apex cannula the collet nut and collet must be changed on the VAD housing as described above. Otherwise, no specific cannula preparation is necessary. Of note, the long, curved ventricular apex cannula seems to fit within the pericardium better than the straight ventricular cannulae. The outflow cannula requires preclotting. The VAD manufacturer describes two techniques for outlet graft preclotting. Method A involves immersion of the arterial graft in 100 ml of nonheparinized blood mixed with 5 mg protamine sulfate and 5000 U topical thrombin. The graft is massaged for 5 minutes by which time a gel should form on the graft. If a gel has not formed by this time the manufacturer recommends that additional thrombin and protamine sulfate be added to the blood and the graft massaged in the resultant solution for an additional 5 minutes. Method B requires that the arterial graft be immersed in 2 U of cryoprecipitate (50 ml/U) and massaged for 5 minutes. The graft is then moved to a new basin that contains 50 ml of thrombin (1000 U/ml). The graft is massaged for an additional 3-4 minutes. Again, a gel should form on the graft.

11

Table 11.3. Thoratec ventricular assist device cannulae*

Ventricular apex inflow cannula		
Blunt tip	no side holes	27 cm long
Extra long, blunt tip	no side holes	29 cm long
Long curved	no side holes	21 cm long
Atrial inflow cannula		
Long (30 cm long)		
Short (25 cm long)		
Arterial outflow cannula		
Short, straight with 14 mm graft	15 cm long	
Long, straight with 14 mm graft	18 cm long	
Long, straight with 18 mm graft	18 cm long	

* Manufactured by Thoratec Laboratories Corp., Pleasanton, CA.

The graft should be flushed with saline to remove any remaining thrombin. We have found that graft immersion causes gel formation on the blood contacting surface. In order to avoid embolization of this gel, saline irrigation is performed. Despite saline irrigation it is difficult to ensure that all microparticles of gel are removed. Furthermore, if the saline flush is applied under pressure it can be forced through the graft interstices thereby increasing graft porosity. We utilize two 60 ml syringes with blunt 18 gauge needles attached. One syringe is filled with topical thrombin (40,000 U in 40 ml, Thrombin-JMI; Gentrac, Inc., Middleton, WI) while the second is filled with cryoprecipitate (2 U = 40 ml). These solutions are sprayed simultaneously onto the external surface of the graft. The resultant gel provides adequate hemostasis while the risk of gel embolism is decreased.

DEVICE IMPLANTATION

Skin exit sites are located and prepared according to the guidelines provided in Chapter 6: Mechanical Ventricular Assistance as a Bridge to Cardiac Transplantation.[3] If an atrial cannula is to be used this cannula is passed through the subcutaneous tunnel before it is inserted into the heart. This can be accomplished with the proprietary trocar, but we find a more snug tunnel results if a Pean clamp is passed through the tunnel into the mediastinum and the cannula grasped and withdrawn through the tunnel. The atrial cannula is inserted into the atrium through the pursestring sutures and the sutures snared and fixed to the body of the cannula. The cannula should be inserted at least 4 cm into the atrium. The single and double line markers on the atrial cannula are located 5 and 6 cm from the tip, respectively.

For left ventricular apex cannulation the manufacturer recommends that twelve pledgetted double-armed sutures be inserted around the apex to form a circle 3-4 cm in diameter. We utilize 2-0 Ethibond sutures (Ethicon, Inc., Somerville, NJ) with handmade pledgets. We find the premade pledgets to be too small and stiff to function effectively. The handmade pledgets are 7 x 12 mm and are cut from a sheet of felt (PTFE felt, 10.2 x 10.2 cm; Meadox Medicals, Inc., Oakland, NJ). The pledgetted sutures are inserted in a horizontal mattress fashion taking deep bites into the myocardium. A core is removed from the left ventricular apex using a circular coring tool (a 10 mm Hancock apico-aortic conduit coring knife works well—Hancock apical coring knife; Medtronic Blood Systems, Inc., Minneapolis, MN). The apical cannula is inserted into the ventricle and the sutures passed through the sewing cuff and tied snugly. After the apex cannula has been inserted into the heart it is passed through the subcutaneous tunnel.

The distal end of the arterial cannula is cut at a bevel and anastomosed to the ascending aorta with a 4-0 Prolene suture. The distal anastomosis is checked for hemostasis by moving the vascular clamp from the aorta to the graft. After hemostasis is achieved the graft is exteriorized. The inlet and outlet cannulae should have at least 1-2 cm of velour exposed at the skin level. Even if the skin-cannula interface were to shift as the patient changes position, adequate tissue ingrowth should occur. The arterial cannula is cut to match the length of the atrial cannula if one is used. The atrial cannula should **never** be cut as the flared end is necessary for connection to the VAD. If a ventricular apex cannula is employed both the apex and arterial cannulae can be cut to ensure that the patient can sit up without kinking the exteriorized segments of these two conduits. Care should be taken to ensure that the cannulae are not cut too short. Approximately 4 cm of nonwire reinforced polyurethane cannulae are required to facilitate connection to the VAD.

The albumin solution is removed from the VAD and the VAD filled with normal saline. Verify proper VAD orientation by checking the arrows on the inlet and outlet valve housing nuts. The collet nuts and collets are slipped onto the cannulae with the threaded and notched ends pointing toward the VAD, respectively. The inlet cannula is connected first. If an atrial cannula is employed the flared end facilitates attachment to the blood pump. The heart is filled or the heart momentarily compressed if CPB is not employed to fill the left atrial cannula. A tubing clamp is applied to the exteriorized segment of cannula below the blood-air interface. Grasp the VAD firmly and force the atrial cannula onto the inlet connector with the tubing clamp. Ensure that the cannula is fully seated into the groove at the base of the inlet connector (Fig. 11.4). Using the back of a forceps seat the collet as far down onto the inlet connector as possible. Slide the collet nut over the collet and tighten firmly by hand. If a left ventricular apex cannula is employed, the leading edge of the cannula is most easily advanced onto the inlet connector by everting the edge of the cannula with three hemostats (Halsted curved mosquito forcep; V. Mueller, Deerfield, IL).

Deairing is best accomplished with a deairing catheter. A 4-0 Prolene pledgetted box stitch is placed in the outflow graft. A #11 scalpel is used to puncture the graft

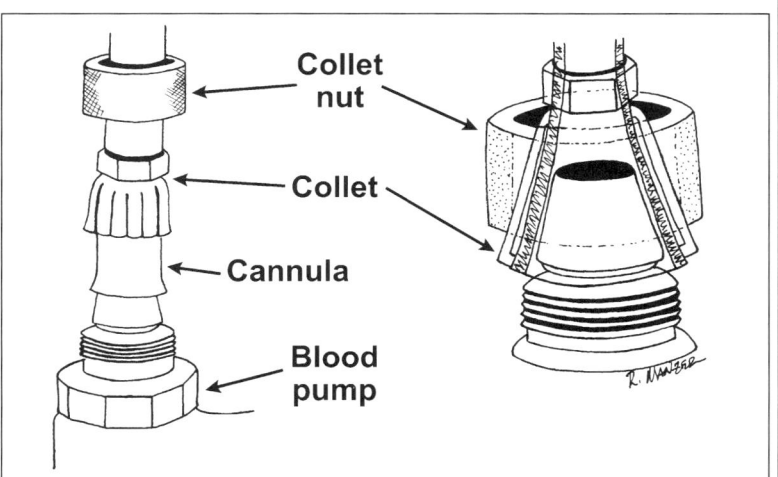

Fig. 11.4. The Thoratec cannula-to-pump connection using the collet and collet nut. The cannula must be completely seated onto the blood pump. When the collet nut is tightened the flange on the collet will hold the cannula in place.

and a Swan-Ganz catheter inserted into the outlet graft and advanced retrograde toward the VAD. The catheter is passed into the VAD via the outflow valve and positioned at the highest portion of the blood sac. The heart and VAD are permitted to fill as the outlet cannula is attached to the outlet connector in a manner identical to that described above. With the patient in the Trendelenburg position, air will tend to rise into the VAD. The heart is shaken and the outlet graft massaged. Air is removed from the VAD by aspirating it through the distal port on the Swan-Ganz catheter. Deairing is facilitated by removing the vascular clamp from the arterial graft. Blood under systemic pressure will pass retrograde across the incompetent outlet valve. When deairing is complete the catheter is withdrawn from the outlet graft and the box stitch tied. Although the manufacturer recommends a 5-7 F right angle angiography catheter for deairing we find the Swan-Ganz catheter works just as well and, more importantly, is readily available in every operating room.

Deairing can also be accomplished without a deairing catheter. In this instance the VAD is first connected to the outlet cannula. The vascular clamp is removed from the arterial graft and a backflow of blood flushes any air in the VAD out through the inflow port. This deairing technique is accomplished by rotating the operating table and VAD itself from side to side to allow the air to rise toward the inlet port. As blood emanates from the inlet port the inflow cannula is connected to the VAD.

The glove finger protector is removed from the Y-connector and the electric and pneumatic leads are attached. The VAD is initially hand pumped until all suture lines have been checked and the heart and VAD have been completely deaired. Until deairing is deemed complete the patient remains in the Trendelenburg position with an aortic root vent in place. When deairing is complete the aortic root vent is removed and the patient returned to a supine position. The pneumatic and electric leads are connected to the drive console and pumping is initiated in the fixed rate mode at a low beat rate.

THERMO CARDIOSYSTEMS HEARTMATE LVAS

DEVICE PREPARATION

The device preparation and implantation technique is essentially the same for pneumatic and electric versions of the Thermo Cardiosystems HeartMate LVAS (Thermo Cardiosystems, Inc., Woburn, MA). The VAD itself does not require any preparation other than to protect the pneumatic/electric connector from exposure to fluids. The inlet and outlet valved conduits contain porcine heterografts which must be washed and the conduits preclotted prior to implantation. The valved conduits are rinsed in three separate basins filled with normal saline. The first rinse is for 10 minutes while the second and third rinses are for 3 minutes each. The valved conduits are gently agitated throughout each rinse cycle. Once the valved conduits have been rinsed they must be preclotted. Although the manufacturers recommend preclotting the valved conduits with nonheparinized blood from the patient, we find this process to be quite time consuming and not particularly thorough. We prefer to spray the external surfaces of the valved conduits with topical thrombin (40,000 U in 40 ml) and cryoprecipitate (2 U = 40 ml). These agents are drawn up in separate 60 ml syringes and sprayed simultaneously onto the external surface of the valved conduits through blunt 18 gauge needles. When a gel forms preclotting is complete. This technique is relatively quick and provides excellent hemostasis. The inflow and outflow valved conduits are inspected to ensure that no debris from the preclotting process has entered the lumen. Excess gel is also wiped from the threaded connectors on the valved conduits. When preclotting is complete the conduits are connected to the inflow and outflow ports of the VAD. Correct orientation is assured by observing the arrows that designate the direction of blood flow on the VAD housing. The VAD is held vertically and filled with normal saline, rotating and agitating the pump to ensure air bubbles have been dislodged. A nonpowdered glove finger is placed over the end of the inflow valved conduit. A solid thread protector is placed over the end of the outflow valved conduit. The primed pump with valved conduits attached is immersed in a basin of saline containing an antibiotic (1 gm vancomycin hydrochloride in 1 L normal saline) taking care to keep the protective bullet on the end of the drive line dry. The drive line adjacent to the VAD should be wrapped tightly with a moist, antibiotic soaked sponge to ensure it is well wetted. Note that drive console preparation, the initial pneumatic vent cycle and controller initialization for the electric VAD have not been described but need to be accomplished according to manufacturer's specifications. One helpful technique to ensure sterility is maintained when preparing the electric VAD is to pass the controller cables off the operative field through a protective sterile sheath. We have found a video camera arm drape (Universal video camera/laser arm drape, 17.5 x 240 cm; Microtele Medical, Inc., Columbus, MS) to serve this purpose well.

The arterial graft portion of the outflow conduit also needs to be preclotted. Remove the metal ring connector from the outflow graft and soak the graft in albumin (20 ml, 25% solution; Immuno-U.S., Inc., Rochester MI). Bake the outflow graft in the autoclave (10 min, 135°C). Once the outflow graft has cooled replace the metal ring connector onto the metal hub of the outflow graft ensuring that the threads are facing away from the graft. The open thread protector is threaded onto the metal ring connector. This protects the metal ring connector and ensures that it is positioned correctly.

DEVICE IMPLANTATION

There are three important decisions that must be made when implanting the Thermo Cardiosystems VAD. The first concerns whether the VAD will be positioned within the abdomen or in a preperitoneal pocket.[4] Secondly, the drive line skin exit site must be determined. Thirdly, a decision must be made whether to implant the device with or without a brief period of cardioplegic arrest. The VAD pocket or intraperitoneal dissection and drive line tunnel are best created before heparin is administered and CPB initiated. We prefer a preperitoneal pocket as it reduces the potential for intraabdominal complications and ensures an earlier return of peristalsis. Proponents of intraabdominal placement note that the implantation is faster as there is less dissection and there is a lower incidence of sepsis as intraabdominal placement is not associated with a large avascular dead space as is the preperitoneal pocket. Traditionally, the drive line passed through a subcutaneous tunnel and exited the skin in the left lower quadrant. Although this allows the drive line to pass in a straight line from the VAD to the skin, the exit site is often below the belt line and can interfere with clothing. More recently, it has been suggested that the drive line be brought across the midline to exit the skin in the right upper quadrant. Motion of the drive line with respect to the skin is decreased as the drive line is more easily fixed using an abdominal binder. The right upper quadrant skin exit site is now preferred. However, we have found it virtually impossible to bring the drive line across the midline in an obese patient as the drive line is too short. It is imperative that the velour covered portion of the drive line be present at the skin exit site to ensure good tissue ingrowth. With respect to the need for a brief period of cardioplegic arrest we believe that it allows us to better inspect the endocardial surface of the heart so that left ventricular thrombus, if present, can be removed in its entirety. Cardioplegic arrest also allows an outlet graft-to-aortic anastomosis in those patients who have a short aorta or who for a hemodynamic reason do not tolerate placement of an aortic partial occlusion clamp.

The outlet graft to aortic anastomosis is accomplished first. If the patient tolerates and their anatomy permits placement of a partial occlusion clamp, this anastomosis is performed before CPB is instituted. The graft is cut to length. If the graft is stretched to length and the metal connector positioned at the xiphoid process the graft will be of correct length if transected at the point at which it abuts the aorta. The graft-to-aortic anastomosis is accomplished with a running 4-0 Prolene suture. The aortic clamp is removed and a vascular clamp applied to the outlet graft. The anastomosis is inspected for hemostasis.

Under cardioplegic arrest the heart is elevated and the core removed from the left ventricular apex using the coring knife provided. We prefer to place the sutures about the left ventricular apex after the core is removed. Pledgetted 2-0 Ethibond sutures are employed using handmade pledgets as described in the section on the Thoratec device. The sutures are placed in a horizontal mattress fashion passing the needle from the epicardial to the endocardial surface of the heart. By so doing, the myocardium is forced to evert slightly as the apex cannula is inserted. The potential for inflow cannula obstruction secondary to an overriding rim of myocardium is eliminated. The sutures are passed through the felt portion of the apical sewing ring. The silicone tube faces outward. When the sutures are tied the sewing ring should be tight against the epicardium. Remove the centering tool.

The VAD is brought to the field and the drive line passed through the subcutaneous tunnel. We find that if the proprietary tunneler is employed it dilates the tunnel to the point where there is no longer tight apposition between the skin and velour surface on the drive line. To achieve a snug tunnel we remove the bullet from the tunneler and pass the thin "stem" of the tunneler through the subcutaneous tunnel. The threaded end of the tunneler is threaded into the end of the white protective bullet located on the distal tip of the drive line. As the tunneler "stem" is withdrawn the beveled end of the white protective bullet creates its own snug subcutaneous tunnel.

The VAD is positioned in the pocket and the inlet conduit passed through the opening in the left hemidiaphragm. The glove finger is removed from the inlet conduit. The leading edge of the silicone tube that is attached to the left ventricular apex is everted with three hemostats. The silicone tube is advanced over the inlet conduit until it abuts the base of the conduit. The suture provided as part of the silicone tube is tied about the inlet conduit. The silicone tube is fixed in place with two tie wraps (Ty-Rap, standard cable ties, 186 x 4.67 mm; Thomas & Betts, Memphis, TN) ensuring that the wraps are oriented at 90° with respect to each other (Fig. 11.5).

The heart and VAD are gently shaken as CPB flow is decreased by approximately 1-2 L/min. The heart and VAD are deaired as blood emanates from the outlet valved conduit. The outlet graft is filled with saline and connected to the threaded connector on the outlet valved conduit. A pledgetted 4-0 Prolene box stitch is placed in the highest portion of the outlet graft and an 18 gauge deairing needle inserted. The VAD is hand pulsed using the hand-crank on the pneumatic console or hand-pump for the pneumatic and electric blood pumps, respectively. The patient is placed in the Trendelenburg position and the VAD connected to the appropriate controller and drive unit. The vascular clamp is removed from the

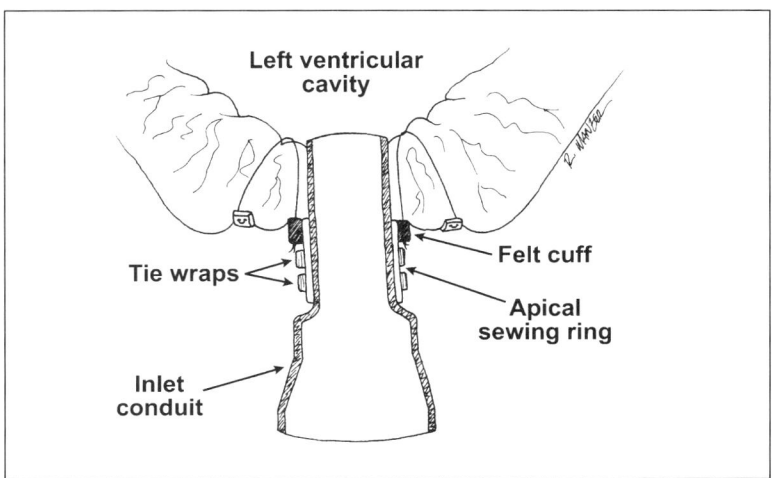

11

Fig. 11.5. The Thermo Cardiosystems LVAS left ventricular apex cannulation technique. Buttressed braided sutures pass through the full thickness of the ventricular apex myocardium. The sutures then pass through the sewing ring on the apical cuff. The inlet cannula is inserted into the apical cuff and held in place with tie wraps.

vascular graft as VAD pumping is initiated. The needle remains in the outlet graft and the vent in the aortic root until deairing is deemed complete. The patient is initially pumped in a fixed rate mode at a low beat rate. When deairing is complete the outlet graft and aortic root vents are removed and the patient returned to a supine position. CPB flow is reduced as VAD flow is serially increased. If VAD filling is satisfactory the VAD can be converted to an automatic mode following the discontinuation of CPB.

NOVACOR N-100 LVAD

DEVICE PREPARATION

Device preparation and implantation techniques for the Novacor N-100 LVAD (Novacor Division, Baxter Healthcare, Corp., Santa Cruz, CA) are in many respects very similar to the techniques employed when implanting the Thermo Cardiosystems HeartMate LVAS. The blood contacting surface of the blood pump is passivated with a 5% albumin solution. The percutaneous vent tube containing power and control leads is protected from fluid exposure during the priming process. The inlet and outlet valved conduits contain porcine heterografts that must be rinsed prior to implantation. The valved conduits and inflow cannula and outflow conduit extension are sealed with an impermeable polymer and do not require preclotting.

DEVICE IMPLANTATION

Experienced Novacor users employ a preperitoneal pocket.[5-7] CPB with or without cardioplegic arrest is employed during device insertion. If the patient tolerates placement of a partial occlusion clamp on the right anterolateral aspect of the ascending aorta, the outlet graft-to-aortic anastomosis is best accomplished prior to the initiation of CPB. The outlet conduit extension is cut to length and the anastomosis between the outlet graft and aorta accomplished with a running row of 4-0 Prolene suture. A vascular clamp is placed on the outlet graft adjacent to the anastomosis as the partial occlusion clamp is removed from the aorta. The anastomosis is checked for hemostasis.

The patient is placed on CPB and normothermia maintained. The ventricular apex is elevated and buttressed 2-0 Ethibond sutures placed in a horizontal mattress fashion about the apical dimple.[6] The sutures are passed through the sewing ring and tied securing the sewing ring to the epicardial surface of the heart. With the patient in the Trendelenburg position a circular ventriculotomy is created in the left ventricular apex with a coring knife.

The blood pump is positioned within the preperitoneal pocket. The drive line is tunneled, caudad to the umbilicus and exits the skin in the right upper quadrant. The apical cannula is passed through the left hemidiaphragm and inserted into the left ventricle. The pursestring suture on the apical sewing ring is tied. The skirt on the apical cannula is sutured circumferentially to the apical sewing ring.

Alternatively, the heart may be arrested after the buttressed sutures are placed about the left ventricular apex.[5]

The valved conduits are attached to the blood pump and the pump filled with saline. CPB flow is reduced and the heart permitted to fill. Blood travels from the native left ventricle through the left VAD and is permitted to emanate from the outlet conduit. The device is grossly deaired with single pump strokes before the outlet graft is connected to the outlet conduit. Final deairing is accomplished through a stab wound or vent inserted into the highest point on the outlet graft. The patient remains in the Trendelenburg position with a vascular clamp on the outlet graft adjacent to the ascending aorta until deairing is complete. The deairing site is repaired as the vascular clamp is removed from the outlet graft. The patient is returned to a supine position as the VAD pumping rate is increased and CPB flow is reduced.

(Acknowledgment: The authors gratefully acknowledge Linda Reed Strauss, Vice President, Novacor Division, Baxter Healthcare Corp., who reviewed the section on the Novacor VAD for accuracy.)

REFERENCES

1. Richenbacher WE, Marks JD. Cannula selection and cannulation techniques for nonpulsatile mechanical ventricular assistance. Artif Organs 1995; 19:519-524.

2. Jett GK. Atrial cannulation for left ventricular assistance: Superiority of the dome approach. Ann Thorac Surg 1996; 61:1014-1015.

3. Ganzel BL, Gray LA Jr, Slater AD et al. Surgical techniques for the implantation of heterotopic prosthetic ventricles. Ann Thorac Surg 1989; 47:113-120.

4. Oz MC, Goldstein DJ, Rose EA. Preperitoneal placement of ventricular assist devices: An illustrated stepwise approach. J Card Surg 1995; 10:288-294.

5. Loisance D, Cooper GJ, Deleuze PH et al. Bridge to transplantation with the wearable Novacor left ventricular assist system: operative technique. Eur J Cardiothorac Surg 1995; 9:95-98.

6. Pennington DG, McBride LR, Swartz MT. Implantation technique for the Novacor left ventricular assist system. J Thorac Cardiovasc Surg 1994; 108:604-608.

7. Scheld HH, Hammel D, Schmid Ch et al. Beating heart implantation of a wearable Novacor left-ventricular assist device. Thorac and Cardiovasc Surg 1996; 44:62-66.

11

Financial Aspects of a Mechanical Circulatory Support Program

Jane E. Reedy

INTRODUCTION

Mechanical circulatory support for patients with cardiogenic shock is rapidly becoming an integral part of cardiac surgery. Ventricular assist devices (VADs) are useful for two major patient groups: those with end stage heart disease who require VAD support as a bridge to heart transplantation while awaiting a suitable donor heart and those who cannot be weaned from cardiopulmonary bypass (CPB) after a cardiac surgical procedure.

There have been many debates about the cost of mechanical circulatory support, often within the context of national healthcare expenditures and what society can afford to provide in terms of medical care for its citizens. VAD support offers obvious benefits but is considered an expensive procedure because of device costs as well as the severe nature of the illnesses they are designed to treat.

The economic burden of heart failure continues to grow annually and heart failure is now the single most expensive healthcare problem in the United States. In 1997, the American Heart Association reported that heart failure affected 4.78 million people in the United States with at least 400,000 new cases diagnosed each year. At least 35% of all patients with heart failure are admitted to the hospital each year.

VAD support is typically used as a last ditch effort to save a patient's life. Thus, in weighing the economic and social costs of end stage heart failure as opposed to VAD support, the following questions must be asked. What costs associated with medical and surgical management of an end stage cardiac disease patient would have been incurred regardless of whether VAD support was performed? If conventional medical and surgical measures do little but delay the inevitable, what economic and social costs would be saved if a vigorous approach such as VAD support had been attempted earlier? Currently, no one has the answers to these questions. Calculating the cost of VAD support without taking into consideration the cost of conventional medical management and failure to achieve patient rehabilitation would not provide a realistic cost benefit or cost effectiveness analysis.

Despite the lack of sophisticated cost analysis and cost benefit data most insurers cover procedures using products that are approved by the Food and Drug Administration (FDA) and are used for an approved indication that is reasonable and necessary for a patient. Many insurers also recognize that physicians are expanding the reasonable and necessary use of VADs. Thus, some insurers may cover VAD support for additional indications based upon patient medical need and demonstrated improved health outcomes that result from the use of a VAD system.

FOOD AND DRUG ADMINISTRATION APPROVAL PROCESS

Considerable progress has been made over the past decade in the development of mechanical circulatory support devices. Funds from the National Heart, Lung and Blood Institute allowed several groups to begin development of VADs during the 1970s. By the end of the decade results of the first clinical trials with VADs were reported. In 1976, the FDA began overseeing clinical trials to ensure that anticipated benefits to the patient and society outweighed the risks. FDA approval was not granted until efficacy and safety of the VAD system were demonstrated. Numerous pulsatile VAD systems underwent lengthy and costly clinical trials and currently five devices have FDA approval (Table 12.1).

Some clinicians consider VAD systems too costly. However, one must realize that due to the complexity of their design it took more than a decade to develop **12** some of the pumps. In addition, companies' expenses increase annually due to the costs associated with preclinical and clinical testing, research and development of second generation devices, ongoing product modification, selling, and general and administrative expenses. Manufacturers of these devices also have rigorous and costly manufacturing and quality assurance requirements because of exposure to potential product liability claims related to the manufacturing, marketing and sale of human medical devices.

JUSTIFICATION

Active cardiac centers acquire one or more VAD systems primarily for the ability to provide patients with state-of-the-art medical procedures that can offer patients the greatest chance for survival. Over the years, VAD systems have saved lives and demonstrated that the capacity to both repair and rest or replace the heart is a distinct advantage. VAD programs can attract patients who would not

Table 12.1. Food and Drug Administration (FDA) approved ventricular assist device systems

Device	FDA approved indication	Pump drive source	Pump placement	Ventricular support	Duration	ICU discharge	Hospital discharge possible
Thoratec VAD System Thoratec Laboratories Corp., Pleasanton, CA)	Bridge to transplantation Postcardiotomy ventricular failure	Pneumatic	Paracorporeal	LVAD RVAD BVAD	Short-Long	Yes	Yes
Abiomed BVS 5000 VAD (Abiomed, Inc., Danvers, MA)	Postcardiotomy ventricular failure	Pneumatic	External	LVAD RVAD BVAD	Short	No	No
Novacor N-100 LVAD (Novacor Division, Baxter Healthcare, Santa Cruz, CA)	Bridge to transplantation	Electric	Implantable	LVAD	Short-Long	Yes	Yes
Thermo Cardiosystems HeartMate IP LVAS (Thermo Cardiosystems, Inc., Woburn, MA)	Bridge to transplantation	Pneumatic	Implantable	LVAD	Short-Long	Yes	Yes
Thermo Cardiosystems HeartMate VE LVAS (Thermo Cardiosystems, Inc., Woburn, MA)	Bridge to transplantation	Electric	Implantable	LVAD	Short-Long	Yes	Yes

LVAD = left ventricular assist device
RVAD = right ventricular assist device
BVAD = biventricular assist device
ICU = intensive care unit

12

normally be referred to some centers. Many of these patients are evaluated for cardiac transplantation and are transplanted with or without the use of a VAD. Other patients referred to assist device programs include patients who undergo high risk cardiac reparative procedures. At smaller cardiac centers these patients may be considered too high risk to undergo elective operation. However, with VAD stand-by these patients, if necessary, can be supported postoperatively until myocardial recovery occurs or until transplantation can be performed. Primary care physicians and referring cardiologists are more likely to refer their patients to centers with the broadest clinical capabilities.

A successful VAD program can augment a cardiac surgical program by increasing the number of open heart procedures or cardiac transplants as well as improving survival. St. Louis University Medical Center, for example, listed 76 patients as United Network for Organ Sharing (UNOS) Status 1 for cardiac transplantation from 1993-1998. Two patients are currently waiting, three expired, and 71 patients were successfully transplanted and discharged from the hospital. Twenty-five of the 71 UNOS Status 1 patients transplanted were supported with a VAD. During the same interval, 34 UNOS Status 2 patients were transplanted from home with two deaths posttransplant. As a result of having a VAD available this center could inform patients as well as referring physicians that once listed as UNOS Status 1 the patient had a 96% chance of receiving a transplant and surviving.

FINANCIAL IMPLICATIONS

The primary costs associated with starting a VAD program are the purchase of capital and disposable equipment, salaries for staff hired specifically for the VAD program and staff training. A hospital can save costs by using existing staff such as clinical specialists, perfusionists or transplant coordinators and by reducing hours required for staff education with on the job training and self study guides.

Major sources of program income are patient care charges which should include VAD and cannulae charges. Centers may also generate revenue by charging a daily fee for circulatory support personnel, billing for a one time console charge at the time of the VAD implant or a daily fee for console use. This billing practice can recover expenditures for capital and disposable equipment.

Indirect sources of income include center recognition, increased referrals, follow-up care revenue and increased usage of ancillary services. The ability to provide patients with state-of-the-art technology may actually help with insurance contract negotiations.

The financial implications of a ventricular assist device program are elusive. Actual cost data reflecting specific resource consumption are lacking. As with most complex medical procedures it is extremely difficult to calculate cost and therefore charges and rates of reimbursement are more commonly used as determinants of success or failure. However, charges differ significantly from the true cost of the service.

An important distinction needs to be made between the cost of VAD components (VAD pump, cannulae and console) and total costs related to treating a patient in postcardiotomy cardiogenic shock or end stage heart failure. Recent studies have shown that even if VAD supported patients remain in the hospital longer than medically treated patients their average daily in-hospital charges can be much less.[1,2] Additionally, those patients bridged with a VAD exhibit a trend towards an improved transplantation rate and a significantly greater rate of discharge from the hospital than medically managed patients.[2,3]

When comparing the cost of different VAD systems one should not only consider the cost of the system, but also the morbidity, mortality, length of stay and total cost of care. Cost comparisons between long-term VAD systems and less expensive short-term assist devices can be misleading because they do not completely account for all of the procedural costs involved in treating the patient during the initial hospitalization period. The following can substantially increase the cost of care with short-term assist devices.

> Operating room costs associated with reoperations and changing pumps due to clots.
>
> Complications associated with early device removal without full myocardial recovery.
>
> Inability to rehabilitate patients, remove intravenous lines or transfer patients from the expensive intensive care unit (ICU) environment.
>
> Switching a patient without myocardial recovery to a long-term VAD system.
>
> Patient transfer to a hospital with a long-term VAD system.

REIMBURSEMENT OVERVIEW

Hospitals with profitable VAD programs typically establish reimbursement teams that may include the surgeon, cardiologist, surgeon's billing staff, hospital admitting/precertification staff, hospital billing staff, hospital medical records/coding staff and hospital financial planning staff. The team is capable of providing a complete profile of indications for use, implantation procedure, patient management and potential complications, alternative procedures, average hospital and physician charges, average length of stay, average ICU days, quality of life improvement, patient profile and others required by insurers who cover the procedure.

Rapidly changing reimbursement policies can significantly impact healthcare providers, especially those introducing breakthrough medical technologies. Typically, the reimbursement rate is determined by the patient's coverage policy and these reimbursement rates may vary according to negotiations between the insurer, hospital and physicians. In conducting any cost pay back analysis a hospital or physician must consider its patient payor mix (Medicare vs. private payer vs. HMO), its geographic location (large urban vs. other urban vs. rural) and its teaching status (teaching vs. community).

To maintain a profitable VAD program hospitals and physicians must have a clear understanding of the patient's coverage policy, appropriate coding for VADs and means to verify adequate payment from the insurers. Centers may also need to renegotiate some contracts to include VAD support.

PATIENT'S COVERAGE POLICY

Most insurers want to provide coverage for the latest and most effective treatments available. However, payment methodologies and payment levels for VAD support can vary considerably. Many private insurers pay hospitals based on the submitted hospital charges or costs. Other payors have established a fixed payment amount in some cases based on a negotiated per diem rate. In other cases, such as with Medicare, payment is based on a prospectively set amount, the Diagnosis Related Group (DRG) payments. Still others may negotiate payment levels for VAD procedures on a patient by patient basis consistent with the terms of the contract between the insurer and hospital.

CODING

Billing staff at the hospital or physician's office place codes on claim forms submitted to insurers to describe procedures performed and patient diseases or products. Insurers typically determine payment level using these codes.

To accurately code one must be very familiar with the coding systems and the reimbursement rules underlying that system. Lack of knowledge can lead to incomplete and inaccurate billing which may lead to denial or inadequate reimbursement. After submitting a bill to a third-party payor all too many people mistakenly assume that the payment received is the maximum allowed. Human and computer errors can happen, therefore, reimbursement tracking is crucial.

All insurers have a process for appealing denied claims. Whenever a claim is returned the reason will be stated. Often, returned claims will simply require additional information, such as a diagnosis or procedure code. The first step in an appeal is to resubmit the denied claim with the requested information.

If the resubmitted claim is not approved the next step is to call the insurer's claims manager or medical director to obtain a review or hearing. Device manufacturers are also willing to help the customer understand why the claim or coverage was denied and assist with the appeal process.

HOSPITAL

The Internal Classification of Diseases, Ninth Revision, Clinical Modification (ICD-9-CM) codes are used to allow the hospital to bill for procedures directly related to assist devices. For VADs, there are two potentially applicable ICD-9-CM procedure codes to describe the implant and others to describe changing or removing the VAD (Table 12.2).

In January 1997, the American Hospital Association ICD-9-CM Editorial Advisory Board decided that the most appropriate ICD-9-CM procedure code for the Thoratec VAD (Thoratec Laboratories, Corp., Pleasanton, CA) system was 37.66. The reimbursement rate has the potential to be higher for Medicare patients when ICD-9-CM procedure code 37.66 is used instead of ICD-9-CM procedure code 37.65 because of the related DRGs.

There are also ICD-9-CM patient diagnosis or disease codes that typically must be included on the claim form from the hospital or physician. These can also be important in the insurer's reimbursement analysis. Hospitals and physicians should refer to the complete listing of ICD-9-CM disease codes to ensure that the code used accurately describes the patient's condition.[4]

PHYSICIAN

The physician reimbursement is based upon the code describing the procedure, the relative value of that procedure and the conversion formula. The Physicians' Current Procedural Terminology (CPT) codes provide the basis for payment as it describes physician provided services (Table 12.3).[5] An ICD-9-CM code is needed to indicate the reason the service was performed or the diagnosis and the Health Care Financing Administration (HCFA) Common Procedural Coding System (HCPCS) may be required to describe what supplies were used.[6] These coding systems standardize the reporting procedures of physicians and form the basis from which reimbursement is determined.

The payment amount to physicians for VAD procedures is based on the insurer's policy with respect to determining physician payment. Some insurers pay on the

12

Table 12.2. Hospital procedure International Classification of Diseases, Ninth Revision, Clinical Modification (ICD-9-CM codes)

Code	Descriptor
37.65	Implant of an external, pulsatile heart assist system
37.66	Implant of an implantable, pulsatile heart assist system
37.63	Replacement and repair of heart assist system
37.64	Removal of heart assist system

Table 12.3. Physicians' current procedural terminology (CPT) codes

Code	Descriptor	RVU
33975	Implantation of ventricular assist device; single ventricular support	21.60
33976	Implantation of ventricular assist device; biventricular support	29.10
33977	Removal of ventricular assist device; single ventricular support	19.29
33978	Removal of ventricular assist device; biventricular support	21.73

RVU = relative value unit

basis of the physician's submitted charge for the procedure. Other insurers established fixed payment amounts for physician services. Under Medicare, for example, physicians are paid according to the Resource Based Relative Value Scale (RBRVS). Each procedure identified by a CPT code has been assigned a relative value unit (Table 12.3) which when multiplied by a conversion factor and geographic index produces the payment level.

Although the private insurance companies function very much like Medicare they are much less consistent and are sensitive to medical marketplace considerations. Relative values for physicians and Medicare variations are the most commonly used rules and regulations in the private sector. In general, each insurance company formulates its own payment policies. Familiarity with these policies should optimize claim collections.

The global fee period is the period of time, including the date of surgery, for which the physician is paid one fee for all care related to that particular procedure. Private payer global days vary. If dealing within a global fee period, modifiers are the only way to be paid for additional services. Reviewing the additional service or procedure is required to determine whether it is covered under the global fee. If not, use the proper modifier to identify the service or procedure to the carrier as a payable service.

The 1999 Medicare RBRVS Final Rule decreased the global billing for VAD insertion from 90 days to 10 days.[7] Therefore, all services provided after 10 days will be separately and legally billable. Knowing the global fee period and billing appropriately and consistently will help alleviate time spent appealing denials.

MEDICARE COVERAGE

Medicare revised its national coverage policy on Artificial Hearts and Related Devices in April 1997. As a result, HCFA broadened Medicare's coverage policy for all VADs that have been approved by the FDA when used in accordance with their approved indications for use (Table 12.4).

Cases are classified into DRGs based on the principle diagnosis, up to eight additional diagnoses and up to six procedures performed during the stay as well as age, sex and discharge status of the patient. This diagnosis and procedure information is reported by using ICD-9-CM codes.

Currently, a DRG does not exist that specifically covers a VAD. For example, a Medicare patient bridged to cardiac transplantation using ICD-9-CM code 37.65 or 37.66 will typically fall into DRG 103 (cardiac transplantation). Patients not bridged to transplantation will fall into other DRG categories (Table 12.5).

To improve reimbursement for implantation of an implantable pulsatile heart assist system for patients not transplanted, the Department of Health and Human Services moved procedure code 37.66 from DRG 108 (other cardiothoracic surgery) to DRG 104 or 105 (Table 12.5). While VADs are still more expensive than the average case in these DRGs, payment would be improved.[8]

Table 12.4. Medicare coverage Issues Manual (CHFA-Publication 6)

65-15 Artificial Hearts and Related Devices

A ventricular assist device (VAD) is used to assist a damaged or weakened heart in pumping blood. VADs are used as either a bridge to a heart transplant or for support of blood circulation postcardiotomy, which is the period following open heart surgery. VADs used for support of blood circulation postcardiotomy are covered only if they have received approval from the FDA for that purpose, and the VADs are used according to the FDA approved labeling instructions. Since there is no authoritative evidence substantiating the safety and effectiveness of VAD used as a replacement for the human heart, Medicare does not cover this device when used as an artificial heart.

All of the following criteria must be fulfilled in order for Medicare coverage to be provided for a VAD used as a bridge to transplant:

The VAD must be used in accordance with the FDA approved labeling instructions. This means that the VAD is used as a temporary mechanical circulatory support for approved transplant candidates as a bridge to cardiac transplantation.

The patient is approved and listed as a candidate for heart transplantation by a Medicare approved heart transplant center.

The VAD is implanted in a Medicare approved heart transplant center on a patient who is listed by that center. If the patient is listed by another Medicare approved transplant center, the implanting center must receive written permission from the center under which the patient is listed. Centers implanting VADs should make every reasonable effort to transplant patients on such devices as soon as medically reasonable. Ideally, the centers should determine patient specific timetables for transplantation and should not maintain such patients on VADs if suitable hearts become available.

REIMBURSEMENT STRATEGIES

Communication is essential between the hospital, clinicians and third-party payors. An individual who does routine approvals according to specific guidelines and policies may deny coverage for VAD support if it is costly and an unfamiliar procedure. It could be helpful to talk to a case manager or medical director about the medical necessity of the proposed procedure.

Preauthorization, if required, should be obtained in writing or written confirmation sent to the insurer. If insurers are not familiar with VAD procedures it can be helpful to provide proof of FDA approval or therapy effectiveness and acceptance in the medical community as well as cost savings that can be realized by transferring patients out of the ICU or discharging patients from the hospital. Other helpful information includes product overview, patient history and medical needs, operative procedure, Medicare coverage policy and articles and references from peer reviewed journals.

Using an FDA approved device and obtaining concurrent transplant and assist approval may maximize reimbursement. Find other practical tips in Table 12.6. It is important to have a clear understanding of the patient's coverage policy and to tailor the reimbursement strategy accordingly.

Centers should also use the coverage decisions of other payors as a benchmark. Also, ask other hospitals that use VADs what their experience with reim-

Table 12.5. Diagnosis related groups (DRGs)

DRGs	Descriptor	Relative Weight 1999
103	Heart transplant	17.79
104	Cardiac valve and other major cardiothoracic procedure with cardiac catheterization	7.28
105	Cardiac valve and other major cardiothoracic procedure without cardiac catheterization	5.71
106	Coronary bypass with percutaneous transluminal coronary angioplasty	7.37
107	Coronary bypass with cardiac catheterization	5.50
109	Coronary bypass without cardiac catheterization	4.07
110	Major cardiovascular procedure with complications and comorbidity	4.15
111	Major cardiovascular procedure without complications and comorbidity	2.22

bursement has been. Find out which companies have reimbursed for the same procedure. In some cases, it may be useful to have the patient's spouse or a family member and their employer contact the insurer directly. Have them pressure the insurer to increase the reimbursement.

Some insurers have contracts with specific centers. If a patient acutely deteriorates and is brought to a noncontracted center for VAD placement the insurance company may deny payment because the patient is "out of network." In this case, the insurer's medical director or case manager should be notified that the patient was acutely deteriorating upon arrival to your center and probably would not have survived the longer trip to the contract center. Present a case that your center can provide the same services as the contracted center and that it would be dangerous to transfer the patient to another center even if comparable technology was available. Whenever possible, communicate with the insurance company based solely upon medical criteria and what your medical team deems to be best for patient survival.

CONCLUSIONS

During the last decade there have been dramatic improvements in the care of those requiring VAD support. The knowledge and experience gained during this time has helped to reduce costs and improve patient survival. Despite the lack of sophisticated cost analysis and cost-benefit data most insurers will cover procedures using products that are FDA approved for an approved indication that is

Table 12.6. Practical tips for reimbursement

Submit accurate and complete claims.

Make sure the submitted claims are consistent with the insurer's policies.

Get copies of insurer's policies in writing.

If a claim is denied, identify specifically why the insurer denied the claim and prepare a response.

Provide scientific articles that demonstrate that the ventricular assist device (VAD) system is safe, effective and that it makes a significant contribution to saving or extending the patient's life.

Work one-on-one with the insurer. Identify and establish a regular contact with the insurer who understands VAD systems and who can facilitate prompt claims processing.

If questions about medical necessity are to be reviewed by the insurer's medical director, have the patient's physician initiate contact to ensure "doctor-to-doctor" communication.

Demonstrate survival rates of patients treated with VADs are superior to a controlled population (Refer to manufacturer's directions for use).

Demonstrate that patients on a VAD(s) are better transplant candidates because they can be rehabilitated while on the device.

Demonstrate cost savings because patients may be rehabilitated quickly, transferred out of the intensive care unit within days and may be discharged from the hospital.

reasonable and necessary for a particular patient. To maintain a profitable VAD program, however, hospitals and physicians must have a clear understanding of the patient's coverage policy, appropriate coding for VADs and the means to verify adequate payment from insurers.

REFERENCES

1. Cloy MJ, Myers TJ, Stutts LA et al. Hospital charges for conventional therapy versus left ventricular assist system therapy in heart transplant patients. ASAIO J 1995; 41:M535-M539.

2. Mehta SM, Aufiero TX, Pae WE Jr et al. Mechanical ventricular assistance: An economical and effective means of treating end-stage heart disease. Ann Thorac Surg 1995; 60:284-290.

3. Gelijns AC, Richards AF, Williams DL et al. Evolving costs of long-term left ventricular assist device implantation. Ann Thorac Surg 1997; 64:1312-1319.

4. International Classification of Diseases, Ninth Revision, Clinical Modification (ICD-9-CM), Health Care Financing Administration.

5. Physicians Current Procedural Terminology, 1999. American Medical Association.

6. Coding & Reimbursement Sourcebook: Medicare and private-party payer reimbursement rules and regulations. Physician Reimbursement Systems Inc., 1998.

7. Medicare: Physician fee schedule (1999 CY): Payment policies and relative value unit adjustments. Federal Register 1998; 63(211):58813-59190.

8. Medicare: Hospital inpatient prospective payment systems and 1999 FY rates. Federal Register 1998; 63(147):40953-41131.

12

Appendix:
Industrial Contact Information

Intraaortic Balloon Pumps

Arrow International, Inc.
2400 Bernville Road
Reading, PA 19605
Tel: (610) 378-0131
     (800) 233-3187
Fax: (610) 374-5360

Datascope Corp.
Cardiac Assist Division
15 Law Drive
Fairfield, NJ 07004
Tel: (800) 777-4222

Extracorporeal Membrane Oxygenation

Medtronic Perfusion Systems
7611 Northland Drive
Brooklyn Park, MN 55428
Tel: (612) 391-9000
Fax: (612) 391-9101
(oxygenator, heat exchanger)

Ventricular Assist Device
Centrifugal Pumps

Bio-Pump:
Medtronic Perfusion Systems
Bio-Medicus Division
9600 West 76th Street
Eden Prairie, MN 55344
Tel: (800) 854-3570
Fax: (800) 477-5467

3M Sarns Centrifugal Pump System:
3M Healthcare
Sarns/CDI
6200 Jackson Road
Ann Arbor, MI 48103
Tel: (800) 521-2818
Fax: (734) 663-7981

Isoflow Centrifugal Pump:
Bard Cardiopulmonary
25 Computer Drive
Haverhill, MA 01832
Tel: (800) 322-2273
Fax: (800) 223-1187

Pneumatic Pumps

Thoratec VAD:
Thoratec Laboratories Corp.
6035 Stoneridge Drive
Pleasanton, CA 94588
Tel: (800) 528-2577
Fax: (925) 847-8574

HeartMate 1000 IP LVAS:
Thermo Cardiosystems, Inc.
470 Wildwood Street
Woburn, MA 01801
Tel: (888) 743-2781
Fax: (781) 933-4476

Abiomed BVS 5000 Biventricular Support System:
Abiomed, Inc.
33 Cherry Hill Drive
Danvers, MA 01923
Tel: (800) 422-8666
Fax: (925) 847-8574

Electric Pumps

HeartMate 1000 VE LVAS:
Thermo Cardiosystems, Inc.
470 Wildwood Street
Woburn, MA 01801
Tel: (888) 743-2781
Fax: (781) 933-4476

Novacor N-100 LVAD:
Baxter Cardiovascular Group
17211 Redhill Avenue
Irvine, CA 92614
Tel: (800) 854-0567
Fax: (800) 422-9329

Index